Environmental Impact of Biofuels

Environmental Impact of Biofuels

Editor

Corbin Tennyson

scitus
academics

Environmental Impact of Biofuels
Edited by **Corbin Tennyson**

ISBN: 978-1-68117-300-9
Library of Congress Control Number: 2016937485

SCITUS Academics LLC,
www.scitusacademics.com
Box No. 4766, 616 Corporate Way,
Suite 2, Valley Cottage,
NY 10989

Notice

Printed in United States of America on Acid Free Paper

Preface

A biofuel is a fuel that is produced through contemporary biological processes, such as agriculture and anaerobic digestion, rather than a fuel produced by geological processes such as those involved in the formation of fossil fuels, such as coal and petroleum, from prehistoric biological matter. Although biofuel is fast emerging as the oil of the future, a number of environmental effects of biofuel are also surfacing associated with its use. These biofuel environmental effects potentially include decrease in greenhouse gas emissions, deforestation, pollution and the rate of biodegradation. Biofuels are only one component of a range of alternatives for mitigating greenhouse gas emissions. Whether using biodiesel is able to lower greenhouse gas emissions as compared to the fossil fuels may depend on many factors. Carbon dioxide is known to be one of the major greenhouse gases. The plant feedstock used in the making biofuels absorbs carbon dioxide from the atmosphere when it grows and once the biomass is converted into biodiesel and burnt as fuel, the energy released is used to power an engine while the carbon dioxide is released back into the atmosphere. There can be grave effects of biodiesel fuel on environment if deforestation and monoculture farming techniques are used to grow biofuel crops. It may damage the ecosystems and biodiversity and increase the emission of climate change gases rather than helping controlling them. To meet the demand for cheap oil from the tropical region, the amount of arable land is being extended in order to increase production at the cost of tropical rainforest. Another of the environmental effects of biodiesel fuels can be seen in its biodegradation rates, which are five times faster than petroleum diesel over a period of 28 days. Biodiesel fuel blends can also accelerate the rate of petroleum diesel degradation through co-metabolism. Toxicity studies have revealed no mortalities and any toxic effects of biodiesel fuel on environment.This book, Environmental Impact of Biofuels, aims to be a comprehensive summary of current biofuels issues and thereby contribute to the understanding of this important topic. This book is recommended for anyone interested in understanding this diverse and developing theme.

Table of Contents

1

CHAPTER

ENVIRONMENTAL IMPACTS OF PRODUCTION OF BIODIESEL AND ITS USE IN TRANSPORTATION SECTOR

Sippy K Chauhan[1] and Anuradha Shukla[1]

[1] Traffic Planning & Environment Division, Central Road Research Institute (CSIR), New Delhi, India

1. INTRODUCTION

The world is presently confronted with the twin crises of fossil fuel depletion and environmental degradation. The search for alternative fuels, which promise a harmonious correlation with sustainable development, energy conservation, efficiency and environmental preservation, has become highly pronounced in the present context. The fuels of bio-origin can provide a feasible solution to this worldwide petroleum crisis. Gasoline and diesel-driven automobiles are the major sources of greenhouse gases (GHG) emission [3 - 5]. Scientists around the world have explored several alternative energy resources like biomass, biogas [6] primary alcohols, vegetable oils and biodiesel. These alternative energy resources are highly environment-friendly but need to be evaluated on case-to-case basis for their advantages, disadvantages and specific applications. Some of these fuels can be used directly, while some others need to be formulated to bring the relevant properties closer to conventional fuels.

Environmental concerns have increased significantly in the world over the past decade, particularly after the Earth Summit-92. Excessive use of fossil fuels has led to global environmental degradation effects such as greenhouse effect, acid rain, ozone depletion and climate change. So there is need to develop or find alternative ways to power the world's motor vehicles.

2 Environmental Impact of Biofuels

There are two global biorenewable liquid transportation fuels that might replace gasoline and diesel fuel. These are bioethanol and biodiesel. Bioethanol is good alternate fuel that is produced almost entirely from food crops. Biodiesel has become more attractive recently because of its environmental benefits.

Transport is one of the main energy consuming sectors. It is assumed that biodiesel is used as a fossil diesel replacement and that bioethanol is used as a gasoline replacement. Biomass based energy sources for heat; electricity and transportation fuels are potentially carbon dioxide neutral recycle the same carbon atoms. Due to its widespread availability, biorenewable fuel technology will potentially employ more people than fossil fuel based technology [7].

The term biofuel is referred to as solid, liquid or gaseous fuels that are predominantly produced from biorenewable or combustible renewable feedstocks [8]. Liquid biofuels are important for the future because they replace petroleum fuels. Biofuels are generally considered as offering many priorities, including sustainability, reduction of greenhouse gas emissions, regional developments, social structure and agriculture, security of supply [9].

The biggest difference between biofuels and petroleum feedstocks is oxygen content. Biofuels are non polluting, locally available, accessible, sustainable and are a reliable fuel obtained from renewable sources. Electricity generation from biofuel has been found to be a promising method in near future. The future of biomass electricity generation lies in biomass integration gasification / gas turbine technology, which offers high energy conversion efficiencies.

First generation biofuels refers to biofuels made from starch, sugar, vegetable oils or animal fats using conventional technology. The basic feedstocks for the production of first generation biofuels are often seeds or grains such as wheat, which yields starch that is fermented into bioethanol, or sunflower seeds, which are pressed to yield vegetable oil that can be used in biodiesel. Table 1 shows the classification of renewable biofuels based on their production technologies [10].

Second and third generation bio fuels are also called advanced bio fuels. Second generation bio fuels are mainly made from non – food crops like wheat straw, corn, wood etc. On the other hand appearing the fourth generation is based in the conversion of vegoil and biodiesel into bio gasoline using the most advanced technology.

Table 1. Classification of renewable biofuels based on their production technologies

Generation	Feedstock	Example
First Generation biofuels	Sugar, Starch, vegetable oils, or animal fats	Bioalcohols, vegetable oil, biodiesel, biogas
Second Generation biofuels	Non food crops, wheat straw, corn, wood, solid waste, energy crops	Bioalcohols, bio – oil, bio – DMF, wood diesel
Third Generation biofuels	Algae	Vegetable oil, biodiesel
Fourth Generation biofuels	Vegetable oil, biodiesel	Biogasoline

Renewable liquids bio fuels for transportation have recently attracted huge attention in different countries all over the world because of its renewability, sustainability, common availability, regional development, rural manufacturing, jobs, reduction of greenhouse gas emissions and its biodegradability. Table 2 shows the availability of modern transportation fuels. There are several reasons for biodiesel to be considered as relevant technologies by both developing and industrialized countries [7]. They include energy security reasons, environmental concerns, foreign exchange savings and socioeconomic issues related to rural sector.

Due to its environmental merits, the share of bio fuel in the automotive fuel market will grow fast in the next decade [11, 12]. The advantages of bio fuels are the following –

a. They are easily available from biomass sources

b. They represent a carbon dioxide cycle in combustion

c. They have a considerable environmentally friendly potential

d. They have many benefits for the environment, economy and consumer and

e. They are biodegradable and contribute to sustainability [13].

Various scenarios have resulted in high estimates of bio fuels in the future energy system. The availability of resources is an important factor. The rationale is to facilitate the transition from the hydrocarbon economy to the carbohydrate economy by using biomass to produce bio ethanol and bio methanol as replacements for traditional oil based fuels and feed stocks.

Table 2. Availability of modern transportation fuels

Fuel Type	Availability	
	Current	Future
Gasoline	Excellent	Moderate poor
Bioethanol	Moderate	Excellent
Biodiesel	Moderate	Excellent
Compressed natural gas (CNG)	Excellent	Moderate
Hydrogen for fuel cells	Poor	Excellent

The refining, transport and combustion of bio fuels can result in significant environmental costs, particularly on local water and air quality. Generally, these effects pale in comparison to those generated by the use of fossil fuels, where the main detrimental environmental effects originate from the vehicle exhaust pipe. Even so, these impacts could expand considerably as bio fuel production increases to meet rapidly rising global demand. However, more sustainable practices and new technologies offer the potential for environmental improvements.

This chapter elaborates the main environmental impacts associated with bio fuels processing, transport and use. In order to provide comparison, it first describes some of the environmental costs resulting from processing and the use of petroleum transport fuels.

2. ENVIRONMENTAL COSTS OF PETROLEUM REFINING AND USE

While the use of oil has brought incalculable benefits to modern industrialised society, it has also extracted great costs, particularly to the local and global environments. Most of these things occur during oil refining and fuel consumption. Delucchi [14] estimated that in the US the costs of environmental externalities associated with oil and motor vehicle use totally between US $ 54 to $234 billion in 1991 alone. Human mortality and disease due to air pollution accounted for more than three quarters of these costs. In Germany it estimated that the quantifiable costs of air pollution and carbon dioxide emissions associated with the transport sector in 1998 totalled about US $ 14.5 billion.

3. OIL REFINING

Refining of petroleum is an energy intensive, water hungry and very highly polluting process. Everyday, average US refinery releases 41, 640 litres of oil and other chemicals into the air, soil and water [15]. Population lives around that location may feel higher incidences of respiratory problems, skin irritation, nausea, eye problems, headaches, birth defects, cancers etc.

Crude oil, chemical inputs and refined products leak from storage tank and spill during transfer points. Numerous toxins are likely to enter the groundwater, including benzene, toluene, ethyl benzene and xylene [16]. Other chemical may split into the air. Gases such as methane and slightly heavier hydrocarbons such as those in gasoline evaporate. Other chemicals enter in the air as combustion products; the most significant of these are sulphur dioxide (SO2), Nitrogen dioxide (NO2), carbon dioxide (CO2), carbon monoxide (CO), dioxins, hydrogen flouride, chlorine, benzene, large and small particulates and lead [17]. It is well established that oil refineries are the largest industrial source of volatile organic compounds and carbon dioxide, which leads to ozone and smog formation in tropospohere. The second known source of sulphur dioxide which contributes to particulate matter and acid rain and the third largest source of nitrogen oxide, all these are known as ozone precursors [18].

4. OIL TRANSPORT

Most of the world's crude oil comes from field far from where it is refined and transported big distances from field to refinery and from refinery to fuel station. Large tanker vessels account for 68% of crude delivery to refineries covering an average of 6600 Km per trip. Oil pipelines, used mainly in places where deliveries can be land based, account for 30% while trucks and train transport the reminder [17].

Invariably oil spills occur along the journey. Although most tanker spillage is relatively minor, while during loading or unloading, even small amounts can damage ecosystems. Pipelines spills, although typically smaller, can also be ecologically disruptive, polluting soil and seeping into ground water. Such spills can be fairly common in region where pipelines are not maintained adequately [19].

Oil is shipped over distances to refineries and from refineries, gasoline and diesel fuels travel via pipelines and trucks to fuel depots. Upon leaving the refinery 59% of refined petroleum fuels enter pipelines before loading to trucks [17]. Gasoline and diesel are lighter hydrocarbons that tend to evaporate, participating in complex reactions that form ozone in the atmosphere. Benzene is another pollutant and is also evaporative and is well known for its carcinogenicity. The most significant hydrological pollutant is methyl tertiary – butyl ether (MTBE), fuel additive derived from petroleum that seeps quickly into nearby groundwater and is a likely carcinogen [20].

5. COMBUSTION OF PETROLEUM FUELS

Compared to bio fuels, petroleum contains a much wider variety of chemical molecules, including far more sulphur. Most of these have been sequestered in the earth for ten and even hundred years of millions years. The burning of gasoline and diesel fuels releases host pollutants and heavy metals that affect local and regional air quality and these are well linked with global warming issues.

Transport related air pollution leads to reduce visibility, damage to vegetation and buildings and increased incidence of human illness and premature death [21]. Road transport is also growing contributor to air pollution in many developing countries / cities particularly where diesel remains the predominant fuel [22]. Table 3 is summarized below shows the main environmental and health impacts associated with the petroleum primary combustion products including CO2, CO, unburned hydrocarbons, NOx, SOx, particulates and in some countries lead [23].

6. ENVIRONMENTAL IMPACTS OF BIOFUEL

Same as like petroleum fuels, bio fuels can have environmental impacts at all stages of their production and use. Relative to fossil fuels, however, the impacts resulting from refining, transporting and using bio fuels are generally significantly smaller. Moreover, there are ways to improve the resource efficiency and impacts of these activities.

Table 3. Environmental and health impacts of emissions from petroleum combustion

Combustion product	Impacts
CO2	Contributes to global warming and climate change
CO	Results from incomplete combustion or burning. In the atmosphere, CO reacts with oxygen to form ozone, a highly reactive molecule that damages plant leaves and human and animal lungs
Benzene	The smallest aromatic hydrocarbon and a highly toxic carcinogen.
NO & NOx	Ozone precursor, they also react with atmospheric water and create acid rain
SO2 & SO3	Acid rain precursors,
Lead	Has been phased out from gasoline in most of the countries, but is still used as an octane enhancer
Particulate matter	Formed from SOx, NOx and hydrocarbons, particulates contribute to ozone formation and affect visibility and hence global warming.

7. WATER USE

Large quantities of water can be utilise for the processing the bio fuel feedstock into fuel. The primary uses of water for biodiesel refining are to wash plants and seeds for processing and then to remove the soap and catalysts from the oils before and final product is shipped out. A typical US Soybean crushing system requires just over 19 Kg water per tonne of oil produced [24]. For each tonne of soybeans that go into the refining process, 170 kg come out as crude de – gummed soybeans oil, 760 kg are soy meal and remaining 70 kg include air and solid and liquid waste [24]. The primary contaminant in wastewater is soybean oil [24].

Production of ethanol, in particular, requires a tremendous amount of water for processing and for evaporative cooling to keep fermentation temperatures at the required level [25]. But some feed stocks are more water intensive than other; each tonne of sugar cane in Brazil, for example, requires as much as 3900 litres for processing [26]. Ethanol processing also results in large volumes of nutrient rich wastewater that, if not cleaned and recycled can speed eutrophication of local rivers and streams by affecting the water's dissolved oxygen content [27]. In addition, sugar mills must be flushed every year, putting huge amount of organic matter into local waterways [28]. In Brazil, 1 litre of ethanol produces about 10 to 15 litres of vinasse which is very hot and corrosive, with a low pH and high mineral content [22]. Today, however, wastewater and vinasse are recycled and used for irrigation and fertilization of Brazil's sugarcane crops, with varying quantities of vinasse used under different conditions as regulated by law [29]. However, some experts caution that vinasse cannot be used where water tables are high, such as in India [30]. Also, if used excessively, vinasse can cause eutrophication of surface water due to the increased load [31]. Filter cake another waste stream from ethanol processing is also recycled as a fertilizer. As a result Brazil has been able to significantly reduce its use of petroleum fertilizers, saving money while creating value from waste products [22].

8. AIR POLLUTION

Among the pollutant that bio refineries emit in to the air are SOx, NOx, VOCs and particulate matter. Emission from corn ethanol plants, for example include SOx, NOx, CO, mercury, particulates and CO [32]. Corn ethanol plants in low a have polluted both water and air, emitting cancer causing chemicals such as formaldehyde and toluene [33].

Biodiesel production require methanol, which has the same environmental cost as those associated with petroleum production. In addition, direct emission from biodiesel processing plants can include air, stream and hexane, which can be used to extract oil from plants and seeds. Hexane is air pollution, and through as much as possible is recovered and recycled, some is emitted into air as well. Sheehan et al (in 1998) estimate that the average US soybean crushing system releases just over 10 Kg of hexane per tonne of oil produced. Alternatives have been found so that hexane is no longer needed; but these options are more costly [34]. In addition, where renewable sources are not used to produce process energy, pollutants associated with the use for natural gas and the generation of steam and electricity are released into air. An estimated 3.6 Kilowatt hours of electricity are required per ton of soybeans entering in a soy biodiesel plant [35]. On the other hand, Fischer-Tropsch [F-T] biodiesel in gasification based and therefore has minimal local air pollution problem [36].

As plant size increases, concerns about pollution – including air emissions, odour releases during the drying of distillers grain in corn ethanol plants and waste water recharges- have risen as well [22]. However, with appropriate regulation and pollution control technologies, emission associated with bio fuels refining can be minimised significantly [37, 38]. For example NOx emission from boiler can be reduced by installing new NOx burner system [39]. VOCs emission, which results primarily from the blending of ethanol with gasoline, can be reduced by mixing fuels at locations where pollutants can be collected and treated [40]. In some cases, new and larger plants are incorporating such emission control system and are finding alternative options that are enable them to reduce such emissions [36].

In addition, much of the air pollution associated with bio fuels refining results from burning of fossils fuels for process heat and power – which in the US, Germany, China and many other countries is mainly coal. Thus, emission can be reduced through traditional power plant control technology or the use of renewably generated power [36].

In Brazil, today, mills and distilleries meet most if not all of their own energy needs with bagasse (a by product of sugar cane crushing), which can generate thermal, mechanical and electrical energy. Some plants even sell surplus electricity into the grid [22]. Elsewhere, agricultural and forestry residues can be used to produce required power and heat; however it is important to ensure that enough residues remain to maintain soil organic matter and nutrient levels [40].

9. BIOFUEL TRANSPORT AND STORAGE

9.1. Water pollution

Pure ethanol and biodiesel fuels offer significant environmental benefits compared to petroleum fuels, making them highly suitable for marine and farm uses, among others. They result in dramatically reduced emission of VOCs and are less toxic to handle the petroleum fuels [41]. One other significant advantage relate specifically to water: both ethanol and biodiesel are biodegradable and break down readily, reducing their potential impact on soil and water [42].

Biodiesel is far more soluble than petroleum diesel, enabling marine animals survive in far higher concentrations of it than petroleum if fuel spills occur (due to lower risk of suffocation) [43]. Such benefits are helping to drive biofuel promotion policies in China, where vehicle have polluted water bodies and gasoline and diesel leakage pipelines has polluted ground water – affecting biodiversity, drinking water and soil resources [25].

At least one study has shown that biodiesel made with rapeseed oil can biodegrade in half the time required for petroleum diesel. Biodiesel also speeds the rate at which biodiesel petroleum blends can biodegrade, which is not the case of ethanol [43]. There is evidence that ethanol's rapid break down deplete the oxygen available in water and soil, actually slowing the breakdown of gasoline. This can increase gasoline's impact on the environment in two ways. First, the harmful chemicals in gasoline persist longer in environment than they otherwise would; benzene, in particular, can last 10 – 150 % longer when gasoline is blended with ethanol. Second, because gasoline breaks down more slowly, it can travel further (up to 2.5 times) in the marine environment, affecting a greater area [44].

Additionally, if ethanol is spilled, it can remobilize gasoline in previously contaminated soils, intensifying the impact of the initial spill. Since up to 85% of such spills occur at gasoline terminals. This is where such a problem is most likely to happen [45]. The transition of high label of ethanol needs to be planned with such impact in mind and should include regulations for the handing for fuels.

9.2. Air pollution

Most biomass is carried to processing plants by truck and most bio fuels are transported by truck as well, although some travel by train or, in Brazil, via pipelines. The environmental impacts associated with transport include the air emission and other pollutants associated with the life cycle of the fuel used – in most cases, petroleum diesel. As demand of bio fuel increases and as consumption exceeds, production in some countries, it is likely that a raising amount of feed stocks and bio fuel will be transported by ship. By shipping is a relatively energy-efficient means of transport, it is also a major source of pollution due primarily to a lack of regulations governing maritime emissions. Pollutants include NOx, SO_2, CO_2, particulate matter and a number of highly toxic substance, such a formaldehyde and poly aromatic hydrocarbons [46]. Emissions from diesel from marine engines represent an ever increasing share of air pollution, and most of these pollutants are released near coastlines, where they can easily be transported over land [46].

The over potential concern associated with bio fuels transport is possibility for spills and evaporation. Bio fuels can leak at the production facility, spill while being transported and leak from above and below ground tanks. They can also evaporate during fuelling and storage and from a vehicle's fuelling system.

In general, 'neat' bio fuels are distinctly less toxic than spills of petroleum fuels. For biodiesel, evaporative emissions are not a particular concern since biodiesel fuel does not have a higher vapour pressure. Neat ethanol has a low rigid vapour pressure (RVP), and when stored as a pure fuels (or even as an E-85 blends), it has a lower vapour pressure than gasoline and thus will have fewer evaporative emissions [36].

The primary concern regarding emission from bio fuel transport has to do with lower- level blends of ethanol in gasoline, which tend to raise vapour pressure of the base gasoline to which ethanol is added. When ethanol is blended up about 40% with gasoline, the two fuel combined have higher evaporative emission than either does on its own. The fuels are mixed via splash blending at the petroleum supply 'rack', so there is a potential for increased evaporative emission from these lower – level blends at the point in the distribution chain and 'downstream'- mainly during vehicles refuelling and from use in the vehicles. These evaporative emissions from a vehicle's fuelling system can increase ozone pollution.

Adding the first few per cent of ethanol generally causes the biggest increase in volatility so increasing the blend level to 2.5 or even 10 per cent will have similar results [47]. Evaporative emissions peak at the blend level between 5 to 10 percent and then start to decline. Once ethanol's share exceeds 40 percent, evaporative VOC emission from the blend are lower than those from gasoline alone [48].

Most international energy agencies (IEA) countries have emission standards requiring the VOC emissions and thus RVP, be controlled [47]. Emission resulting from higher vapour pressure can be controlled by requiring refiners to use base gasoline stock with a lower vapour pressure when blending with ethanol, although this increases cost and reduces production lavels. The US state of California and US federal reformulated gasoline programmes have set caps on vapour pressure that take effect during high ozone seasons in areas that do not meet ambient air quality standard of ozone. As a result, the addition of ethanol does not increase the vapour pressure of the gasoline available during summer months [49]. Emission from permeation are more difficult to control in the on-road fleet, although expert believes that most can be controlled in new vehicles that much meet stricter evaporative emission control standard (such as California LEV 2 and US Federal Tier II), with higher-quality tubes, hoses and other connectors [49].

10. BIOFUEL COMBUSTION

The level of exhaust emissions that results from the burning of ethanol and biodiesel depends upon the fuel (e.g. feedstock and blend), vehicles technology, vehicle tuning and driving cycles [22]. Most studies agree that using bio fuels can significantly reduced most pollutants compared to petroleum fuels, including reductions in controlled pollutant as well as toxic emissions [47]. NOx emissions have been found to increase slightly as blend level rise, although the levels of emissions differ from study to study.

10.1. Ethanol

Ethanol contains no sulphur, olefins, benzene and other aromatics [22]. All of which are component of gasoline that can affect air quality and threaten human health [22]. Benzene is carcinogen, while olefins and some other aromatics are precursor to ground-level Ozone (smog) [47]. Ethanol-gasoline blends also reduce toxic emissions of 1,3-butadiene, toluene, xylene, while few studies have a looked at the impacts on the pollution levels from high blends.

With ethanol fuel combustion, emission of the toxic air pollutants acetaldehyde, formaldehyde, peroxyacetyl nitrate (PAN) increase relative to straight gasoline [47]. Most is emitted as acetaldehyde, a less reactive and less toxic pollutant than formaldehyde. Neither pollutant present in fuel; they are created as by product of incomplete combustion. PAN is an eye irritant that is harmful to plants, is also formed as by product [47]. A US auto-oil industry study determined that combustion of E85 resulted in a slight increase in hydrocarbon emission relative to California reformulated gasoline. It is also found that toxic emission rose as much as two to three folds compared to conventional gasoline, due mainly to an increase in aldehyde emission.

There is concern that aldehydes might be carcinogenic; but the pollutant that are reduced by blending with ethanol (including benzene, 1,3 – butadiene, toluene and xylene) are considered more dangerous for human health. A study done in California determined that acetaldehyde and PAN concentrations increases only slightly with ethanol blends, and a Canadian study concluded the risks of increased aldehyde pollutants are negligible [90]. Because of reactivity of aldehydes, emissions can generally be managed with emission controls [36]. For example, three way catalysts can efficiently minimize aldehyde emissions [50].

Ethanol blended gasoline increase fuel oxygen content, making hydrocarbons in the fuel burn more completely in older vehicles, in particular, thus reducing emission of CO and hydrocarbon emissions [22]. Ethanol used an additive or oxygenate (e.g. 10 percent blend) has been found to achieve CO reductions of 25 percent or more in older vehicles [22]. In fact, one of the goals driving the use of ethanol in US during the 1990s was to reduce hydrocarbons and CO emissions particularly in winter when emission of these pollutants tend to be higher. Ethanol in higher blend will be positively affecting the efficiency of catalytic convertors because of the dilution of Sulphur [51]. Ethanol can be used to make ethyl tertiary butyl eather (ETBE), which is less volatile than ethanol and widely used in the European Union (EU) [51].

As a result of its national ethanol programme *proalcool*, Brazil was one of the first countries in the world to eliminate lead entirely from gasoline. According to Sao Poulo State Environmental Agencies (CETESB), ambient lead concentrations in the Sao Poulo metropolitan region declined from 1.4 gram per cubic meter in 1978 to less than 0.1 gram per cubic meter in 1991 [52]. Most of the countries however have been able to eliminate lead through other means, including a reduction in unnecessarily high octane grades and the development of the chapter refining alternatives (e.g. reforming and isomerisation) [22].

Ethanol use has resulted in significant reductions in other air pollutants as well. Emission of toxic hydrocarbons such as benzene has declined in Brazil, in addition to the emission of sulphur and CO. For Example, Brazil transport related CO emission declined from more than 50grams per kilometre in 1980 to less than 1gram per kilometre in 2000 due to ethanol use. CETESB estimates that urban air pollution in Brazil could be reduced in additional 20-40 percent if the entire vehicle fleet were fuelled by alcohol [22]. In 1998 Denver, Colorado, became the first US city to require blending of gasoline with ethanol; it is used in winter to improve fuel combustion and to reduce CO emissions. As a result, it is estimated that CO level have declined by 50 percent [53].

There is some evidence that emission reductions associated with using ethanol blends, compared to straight gasoline, are not as significant in the cleanest vehicles available today. Durbin et al (2006) tested vehicles that qualified as low emission and ultra low emission in California, and found that emission of non methane hydrocarbons increased as engine temperature rose and that benzene emission increased with higher concentrations of ethanol, while fuel efficiency declined. However, CO emission decreased somewhat with ethanol use [54]. Some of the findings were inconsistence with those of the studies, highlighting the need for further research [54].

As discussed earlier, ethanol used as an oxygenate can reduce emission of several pollutants particularly in older vehicles. However the use of oxygenates such as ethanol (and biodiesel), to alter the fuel to oxygen ratio will not necessarily have a positive effect on emission if a vehicle's air-to-fuel ratio is set low or if too much ethanol is added to gasoline in a vehicle with a fixed air-to-fuel ratio. If that is the case, oxygenate can increase NOx emissions and cause 'lean misfire' increasing hydrocarbon emissions [22]. In fact, Tyson et al (1993) argue that ethanol has no emission related advantages over reformulated gasoline other than the reduction of CO_2 [55].

Ethanol blended with diesel can provide substantial air quality benefits, Blends of 10-15 per cent ethanol (combined with performance additive) result is significantly lower emission compared with pure diesel fuel; exhaust emissions of PM, CO, and NOx decline. For high blends, the results are mixed. Some studies have found higher average CO and hydrocarbon emissions and other have seen reductions in these pollutants. However, all studies, to date, have seen significant decrease in both PM and NOx [47].

Flexible-fuel vehicles (FFV)- which can take virtually any ethanol- gasoline blend up to 85 percent in the US and up to 100 percent in Brazil- are widely used in the Brazil and are becoming increasingly available in the US. However, tests to date have found that the use of FFVs results in higher air emission than new gasoline vehicles [36]. Because it is not possible to tune the combustion controls of vehicles so that it is optimized for all conditions, controls are compromised somewhat to allow for different mixes [56, 57]. It is possible that vehicles dedicated to specific blends, operated on those blends level, would achieve lower emissions than conventional vehicles.

10.2. Biodiesel

Biodiesel – whether pure or blended- results in lower emissions of most pollutants relative to diesel, including significantly lower emission of particulates, sulphur, hydrocarbons, CO, toxins [57]. Emissions vary with engine design, condition of vehicles and quality of fuel. In biodiesel- diesel blends, potential reductions of most pollutants increase almost linearly as the share of biodiesel increases, with the exception of NOx emission [47].

In one of the most comprehensive analyses to date, a US Environmental Protection Agency (EPA) study of biodiesel determined that the impacts on emissions vary depending upon type (feedstock) of biodiesel and the type of petroleum diesel that it is mixed with. Overall animal based biodiesel did better in the study than plant based biodiesel with regard to reducing emission of NOx, CO and particulates. On average, the EPA determined that B20 (made with soybeans) increase NOx emission the least, followed by rapeseed biodiesel and that soybean based biodiesel; the same relationship held true for CO reduction, as well. Reductions in particulate emissions were also greatest for animal based biodiesel [58].

The test carried out by the EPA showed that, when compared with conventional diesel, pure diesel (produced with Soybean oil) resulted an average reduction of particulate matter by 40 percent, CO by 44 per cent, unburned hydrocarbons by 68 percent, polycyclic hydrocarbons (PAHs) by 80 percent, carcinogenic nitrate by 90 percent, sulphate by 100 percent [59].

During 2000, biodiesel become the first alternative of fuel to successfully complete testing for tier 1 and 2 for health effect under the US Clean Air Act. Test determined that, with the exception of minor damage to the lung tissue at high level of exposure, animal observed in the study suffered non biological significant short term effect associated with biodiesel [22].

A 1999 Swedish study by Pedersen et al found that biodiesel (rapeseed methyl ester, or RME) led to an up to tenfold increase in emission of benzene and Ozone precursors compared with Swedish low sulphur diesel fuel, called MKI [60]. However, this study was conducted using a very small reactor; many US and European researches were sceptical about transferring results from this study to the real world for combustion in a diesel engine. Since then, other studies have produced results. For example, Krahl et al (2000) compared 100 percent RME to MKI, fossil diesel fuel and another low sulphur diesel fuel (with high aromatic compounds content and flatter boiling characteristics, known as DF05), using modern DaimlerChrysler diesel engine such as those generally installed in light duty transport vehicles. They concluded that RME lead to significant reduction in CO, hydrocarbons, (HCs), aromatics HCs (including Benzene) and aldehydes, ketones (which contribute to the formation of summer smog) compared with the other fuels [61].

11. IMPACT OF NOX EMISSIONS

Most studies conclude that ethanol and biodiesel emit higher amounts of nitrogen oxides (NOx) than do conventional fuels, even as other emissions decline [47] there are exceptions, however. When ethanol is blended with diesel, NOx, emissions decline relative to pure diesel fuel; and some tropical oils are saturated enough- thus have a high enough cetane value – that they increase NOx less (and in the case of highly saturated oils such as coconut, actually decrease NOx) relative to diesel [62]. NOx are precursor to ground level ozone (smog). In addition, NOx emission increase acid rain and are precursor to fine particulate emissions; associated with health impact include lung tissue damage, reduction in lung function and premature health [63].

The level of NOx emissions found varies significantly from study to study. Some cities, particularly in the US state of California, have complained that ethanol has increased local problems with NOx and ozone [22]. California is using ethanol as an oxygenate meet requirements under US Clean Air Act because concern about water contamination led to the state to ban MTBE. More recently concern about evaporative VOCs emission and combustion emissions of NOx led California to sue the US EPA twice for a waiver; both times the waiver was denied [58]. But both the EPA and California Air Resources Board agreed during the process that ethanol increases NOx slightly in the on-road fleet [64].

Fulton et al (2004), on the other hand, report that the impact of bio fuels on NOx emissions level are relatively minor and can actually be higher or lower than conventional fuel, depending upon the conditions. In fact, there is evidence that NOx level from low ethanol blends range from a 10 percent decrease to a 5 percent increase relative to pure gasoline emission [47].

Studies by US National Renewable Energy Laboratory (NREL) show inconsistent results with regard to biodiesel and NOx, depending upon whether vehicle is driven on the road or in the laboratory. According to McCormick (2005), they have seen 'Nox reductions for testing of vehicles (chassis dyno) and Nox increases for testing of engines (engine dyno). The former, which involves driving an entire car on rollers rather than testing emissions directly from an engine removed from the vehicle, is considered more realistic than the latter [65].

NREL studies of in-use diesel buses have found a statistical significant reduction in NOx emissions with biodiesel. A US auto-oil industry six year collaborative study examined the impact of E85 on exhaust emissions and found that NOx emission were reduced by upto 50 percent relative to conventional gasoline [66]. But India's Central Pollution Control Board has determined that burning biodiesel is a conventional diesel engine increases NOx emissions by about 13 % [57].

Fortunately, newer vehicles designed to meet strict air standards, such as those in California, have very efficient catalyst system that can reduce VOC, NOx and CO emissions from ethanol-gasoline blends to very low levels [36]. With biodiesel, NOx increases can be minimized by optimizing the vehicle engine for the specific blend that will be used [47]. Emission can also be reduced with additives that enhance the cetane value or by using biodiesel made from feedstock with more saturated fats (e.g. tallow is better than canola, which is better than soy) [65].

It is possible to control diesel exhaust using catalysts and particulate filters. High efficiency Diesel Particulate Filter (DPF) remove particulate matter (PM) by filtering engine exhaust; such system can reduce PM emissions by 80 percent or more. However, because of concern about increased oil film dilution during the post- injections. German car manufactures do not accept neat biodiesel in DPF equipped vehicles [67]. There is also concern that the extra injection used to increase emission temperatures for regeneration of the particulate trap result in a dilution of engine oil when RME is used as a fuel, and this dilution can increase engine wear [68]. Rust particles filters, which are available in many new diesel automobiles and significantly reduce emissions of fine particulates, cannot operate with biodiesel [69].

According to some sources, biodiesel do not meet European air emissions standards that went into effect in January 2006 [69], although the Association of German Biofuel Industry noted that biodiesel can meet updated European standards for trucks and commercial vehicles.

Several groups are in the process of developing additives to address the issue of NOx, emissions associated with biodiesel blends, including NREL, the US National Biodiesel Board, the US Department of Agriculture and World Energy Alternatives [65].

12. ADVANCED TECHNOLOGIES

In general, the air quality benefits of biofuel are greater in developing countries, where vehicle emission standards are non-existent or less stringent and where older more polluting cars are more common [70]. For example, the use of ethanol can effectively reduce emissions from CO and hydrocarbons in old technology vehicle today [22]. Less understood, however, are the impacts that biodiesel might have on exhaust emission from vehicles that are underpowered, over-fuelled, overloaded and not well maintained- vehicles that are most prevalent in the world's developing nation [22].

Advances in pollution control technologies for petroleum-fuelled vehicles will reduce. If not eliminate, the relative benefits of biofuels. Greene et al (2004) note that the main benefit of biofuels in such advanced vehicles may be to make it easier to comply with emission standard in the future, thus reducing the cost emission control technologies [36].

At the same time, new technologies are on the horizon. For example, Volkswagen and Daimler Chrysler have invested in biomass-to-liquid (BTL) technologies that convert lignocellulosic fibers into synthetic biodiesel. This process enables them to produce a cleaner burning biofuel. In the future, they hope to optimize fuels and vehicle engines in parallel.

13. CONCLUSION

The refining, transport and combustion of biofuels have environmental costs, particularly on local water and air quality, and these impacts could rise considerably as biofuel production increases to meet rapidly rising global demand. At the same time, more sustainable practices and new technologies offer the potential for environmental improvements.

Increasing efficiencies in water and energy use at refineries can help to reduce both air and water pollution. The UK-based biodiesel producer D1 Oils now recycles both water and methanol used in its refineries and uses biodiesel to run its facilities [71]. Standards and regulations are also needed to minimize pollutants. In addition, encouraging smaller scale distributed facilities will make it easier for communities to manage wastes, while possibly relying on local and more varied feed stocks for bio fuel production and thereby benefiting local economies and farmers.

The combustion of bio fuels- whether blended with conventional fuels or pure-generally results in far local emissions of CO, hydrocarbons, SO_2 and particulate matter (and, in some instances lead) than does the combustion of petroleum fuels. Thus, the use of bio fuels, particularly in order vehicles, can significantly reduce local and regional air pollution, acid deposition and associated health problems; such as asthma, heart and lung disease and cancer (http://www.d1plc.com).

However, the air quality benefits of bio fuels relative to petroleum fuels will diminish as fuel standards and vehicle technologies continue to improve in the industrialised and developing worlds. Even today, the newest vehicle technologies continue to improve in the industrialised and developing worlds. Even today, the newest vehicles available for purchase largely eliminate the release of air pollutants (aside from CO_2) [72]. At the same time, concern about level of NOx and VOC emissions from bio fuels will probably diminish with improvements in vehicles and changes in fuel blends and additives. A combination of next generation bio fuels can make a major contribution to reducing air pollution in the transport sector.

In the developing world, ethanol should be used to replace lead, benzene and other harmful additives required for older cars and because of high blends or pure bio fuels pose minimal air emissions problems and are less harmful to water bodies than petroleum fuels, for all countries it is important to transition these high blends as rapidly as possible, particularly for road transport in highly polluted urban areas and few water transport, wherever feasible.

REFERENCES

1. The Earth has warmed by 0.6º C over the past 30 years and by 0.8º C over the past 100 years, according to US National Aeronautics and Space Administration 2006
2. Kevin A., Baumert, Timothy Herzog and Jonathan Pershing 2005Working Group III (2001)
3. Kesse D G2000Global warming- facts, assessment, countermeasures. J. Petrol Sci Eng 26141149Doi:S0920- 4105 (00)00030- 9
4. X. Cao, 2003Climate change and energy development: implications for developing countries. Resource Policy 296167
5. Y. Johansson, S. Mc Carthy, 1999Global warming post Kyoto: continuing impasse or prospects for progress? Energy Dev Rep Energy, 6971
6. J. D. Murphy, K. Mc Carthy, 2005The optimal production of biogas for use as atransport fuel in Ireland. Renew Energy 3021112127
7. A. Demirbas, 2006Global biofuel strategies, Energy Edu Sci Technol, 172763
8. A. Demiras, 2007Gasoline and diesel fuel blends with alcohols, Energy Edu Sci Technol 198792
9. L. Reijnders, 2006Conditions for the sustainability of biomass based fuel use. Energy Policy 34863876
10. A. Demirbas, 2009Biofuels, Green Energy & Technology
11. S. Kim, B. E. Dale, 2005Life cycle assessment of various cropping systems utilized for producing biofuels: Bioethanol and biodiesel, Biomass Bioenergy 29426439

12. M. F. Demiras, M. Balat, 2006Recent advances on the production and utilization trends of biofuels: A global perspective, Energy Convers Mgmt 4723712381

13. D. Puppan, 2002Environmental evaluation of biofuels, Periodica Polytechnica Ser Soc Man Sci, 2002; 1095116

14. Delucchi, M A1995Summary of non monetary externalities of motor vehicle use, report 9 in series, The Annualized Social Cost of Motor Vehicle Use in US, based on 19901991Data: Summary of Theory, Methods and Datadraft report prepared for Union of Concerned scientists (UCS), Davis, California, Institute of Transportation Studies.

15. EC2003External Costs: Research Results on socio- environmental damage due to electricity and transport, Brussels, EC

16. Sherertz, P C1998Petroleum products, Richmond, VA, Virginia Department of Health, 30 June

17. M. Mann, P. And, Spath, 1997Life cycle assessment of a biomass Gasification Combined cycle Gasification System, Golden, CO, National Renewable Energy Laboratorty

18. D. Doniger, 2001Oil Companies, making record profits, seek Environmental Rollbacks, New York, Natural resources Defense Council, 8 May.

19. I. Bazilescu, B. Lyhus, 1997Russia Oil Spill, Trade and Environment Database case studies, Washington DC

20. Pennsylvania Department of Environmental Protection 2005 Automobile Emissions

21. Abt associates Inc2000The Partculate Related Health Benefits of Reducing Power Plant Emissions, Bethesda, M D, Abt associates Inc,

22. M. Kojima, T. Johson, 2005Potential for Biofuels for Transport in Developing countries, Washington. DC, World Bank.

23. S. Mc Millen, et.al, 2005Biodiesel: Fuel for Thought, Fuel for Connecticu's Futures, Stores, CT, Connecticut Centre for Economic Analysis, University of Connecticut.

24. J. Sheehan, et.al, 1998Life Cycle Inventory of biodiesel and Petroleum Diesel for use in an Urban Bus, Golden, CO, National Renewable Energy Laboratory

25. W. Gehua, et.al, 2005Liquid Biofuels for transportation: Chinese Potential and Implications for sustainable Agriculture and Energy in the 21st Century, Report prepared for GTZ, Beijing

26. M. Peplow, 2005Ethanol production harms environment, researchers claims, Nature, July

27. M. Tampeir, et.al, 2004Identifying Environmentally Preferable Uses for Biomass Resources, Stage 2 Report: Life cycle GHG Emissions Reduction Benefits of selected Feedstock- to- Product Threads, Prepared for Natural Resources Canada and National Research Council Canada, North Vancouver.

28. J. Clay, 2004World Agriculture and Environment, Washington, DC, Island press.

29. I. C. Macedo, et.al, 2005Sugar Cane's Energy: Twelve studies on Brazilian Sugar cane Agribusiness and its Sustainability, Sao Paulo.

30. S. Hunt, J. Sawin, P. with, Stair, 2006Cultivative renewable alternatives to oil, in Worldwatch Institute (ed) State of the World 2006, New York.

31. Sierra Club of Canada2004Mackenzie Pipeline to Fuel Americas' Gas Tank, Press release, Ottawa

32. Environment News Services2005Illinois Ethanol Producer must install Air pollution controls, December

33. P. Beeman, 2005Ethanol plants among Iowa's polluters', Des Moines Register, 11 September

34. J. Core, 2005new method simplifies biodiesel production', Agricultural Research, April

35. j. Sheehan, et al.1998cLife Cycle Inventory of Biodiesel and Petroleum Diesel for Use in an Urban Bus, Golden, CO, National Renewable Energy Laboratory, May

36. Greene et al2004Growing Energy; How Biofuels Can Help End America's Oil Dependence, Washington, DC, Natural Resources Defence Council, December

37. P. Beeman, 2005Ethanol plants among Iowa's polluters', Des Moines Register, 11 September; USEPA (2005b) 'Ethanol Plant Clean Air Act Enforcement Initiative, Washington, DC, updated 1 September 2005,

38. Greene et al2004Growing Energy; How Biofuels Can Help End America's Oil Dependence, Washington, DC, Natural Resources Defence Council, December

39. T. Doggett, 2006EPA seeks to ease US ethanol plants pollution rules', Reuters, 3 March

40. Novozymes and BBI International2005Fuel Ethanol: A Technological Evolution, Grand Forks, ND, June

41. TaTEDO (Tanzania Traditional Energy Development and Environmental Organization)2005Biofuels for transportation in Tanzania: Potential and Implications for Sustainable Agriculture and Energy in 21st Century, Prepared for the Deutsche Gesellschaft fur Technische Zusammenarbeit (GTZ) GmbH, Dar es salaam, September

42. R. D. Perlack, et.al, 1995Biomass fuel from woody crops for electric power generation, Oak Ridge, TN,

43. J. Cook, J. Beyea, analysis. An, environmental. of, of. impacts, crops. energy, the. U. S. A. in, conclusions. methodologies, Washington. D. C. recommendations, 2005

44. Northeast States of Coordinated Air Use Management2001Health, Environmental, Economic Impacts of Adding Ethanol to Gasoline in the Northeast State, 3Water Resources and Associated Health impacts, Lowell, MA, New England Interstate Water Pollution Control Commission cited in Greene et al (2004) Growing Energy; How Biofuels Can Help End America's Oil Dependence, Washington, DC, Natural Resources Defence Council, December; TERI (2005) Liquid Biofuels for Transportation: India Country Study on Potential and Implications for Sustainable Agriculture and Energy, Report prepared for the Deutsche Gesellschaft fur Technische Zusammenarbeit (GTZ) GmbH, New Delhi, October

45. wedel. R. Von, 1999Technical Handbook for Marine Biodiesel in Recreational Boats, second addition, Point Richmond, CA, CytoCalture International,

46. Northeast States of Coordinated Air Use Management2001Health, Environmental, Economic Impacts of Adding Ethanol to Gasoline in the Northeast State, 3Water Resources and Associated Health impacts, Lowell, MA, New England Interstate Water Pollution Control Commission

47. Clean Air Task Force et al2005Prevention of Air pollution from Ships: Reducing Shipping Emissions of Air pollution- Feasible and cost-effective Options submitted by Friends of the Earth International to the Marine Environment Protection Committee, International Maritime Organisation, 7 April.

48. L. Fulton, T. Howes, J. Hardy, 2004Biofuel for transport: An International Perspective, Paris, France, International Energy Agency

49. Northeast States of Coordinated Air Use Management2001Health, Environmental, Economic Impacts of Adding Ethanol to Gasoline in the Northeast State, 3Water Resources and Associated

50. Steven J. Brisby, Manager Fuels Section, Stationary, stationary Source Division, California Air Resources Board e-mail to Janet Sawin, world watch Institute, 1 March 2006.

51. C. Prakash, 1998Use of Higher than 10 volume Percent Ethanol/Gasoline Blend in Gasoline Powered Vehicles, Ottawa, Environment Canada,

52. C. Hammel-Smith, al. Et, 2002Issues Associated with the Use of Higher Ethanol Blends (E17E-24) Golden, CO, NREL, October

53. CETESB (Sao Paulo State Environmental Agency)2003Relatorio de Qualidade do ar no Estado de Sao Paulo, CETESB, Sao Paulo

54. F.O.Licht's World Ethanol and Biofuels Report (2005Denver winter ethanol to stay until 2008' 19December

55. T. Durbin, al. Et, 2006Final Report: Effect of Ethanol and Volatility Parameter on Exhaust Emission, CRC Project E-67

56. K. S. Tyson, C. J. Riley, K. K. Humphreys, 1993Fuel Cycle Evaluations of Biomass Ethanol and Reformulated Gasoline, Golden, CO, NREL

57. J. Earley, T. Earley, M. Straub, 2005Specific Environmental effects of trade liberalization: oil seeds, Washington, DC, International Policy council for food & Agriculture

58. TERI2005Liquid Biofuels for Transportation: India Country Study on Potential and Implications for Sustainable Agriculture and Energy, Report prepared for the Deutsche Gesellschaft fur Technische Zusammenarbeit (GTZ) GmbH, New Delhi, October

59. USEPA2005EPA upholds Reformulated Gas Requirement in California, New York, and Connecticut, Press release, Washington, DC, 2 June

60. K. Becker, G. Francis, 2005Biodiesel from Jatropha Plantations on Degraded Land, Stuttgart, University of Hohenheim cited in TERI (2005) Liquid Biofuels for Transportation: India Country Study on Potential and Implications for Sustainable Agriculture and Energy, Report prepared for the Deutsche Gesellschaft fur Technische Zusammenarbeit (GTZ) GmbH, New Delhi, October

61. J. R. Perdersen, A. Ingemarsson, J. O. Olsson, 1999Oxidation of rapeseed oil, rapeseed methyl ester (RME) and diesel fuel studied with GC/MS', Chemosphere, 381124672474

62. J. Krahl, et al.2001Comparison of Biodiesel with Different Diesel Fuels Regarding Exhaust Gas Emissions and Health Effects, Perdersen, J. R., A. Ingemarsson and J. O. Olsson (1999) 'Oxidation of rapeseed oil, rapeseed methyl ester (RME) and diesel fuel studied with GC/MS', Chemosphere, 381124672474

63. M. Liezzel, Pascual and Raymond R. Tan (2004

64. US EPA (no date) Health and Environmental Impacts of NOx

65. Ronald Hwang, Vehicle policy director, Natural Resources Defense Council, San Francisco, CA, e-mail to Janet Sawin, Worldwatch Institute, 12 March 2006

66. R. Mc Cormick, 2005Effects of Biodiesel on NOx emissions', Presentation to ARB Biodiesel workgroup, National Renewable Energy Laboratory, Golden, CO, 8 June email to Peter Stair, Worldwatch Institute, 20 july 2005

67. AQIRP (Auto/Oil Air /quality /improvement Research Program)1997Program Final Report, AQIRP, January

68. Diesel Net News, Maycited in AMFI Newsletter (Advanced Motor Fuels Information) (2005October,

69. Miljofordon Newsletter no 42005, cited in AMFI Newsletter (Advanced Motor Fuels Information) (2005) October,

70. S. Sinico, 2005Fill it up with Natural, dw-world.de, 22 September; Association of the German Biofuels Industry from Karin Retzlaff of the Association of the German Biofuels Industry, cited in Sinico, S. (2005) Fill it up with Natural, dw-world.de, 22 September

71. Countries. Developing, Fulton. L. in, 2004aDriving ahead: Biofuels for transport around the world', Renewable Energy World, July-August, 180189

72. Emissions, from. Reductions, L. Fulton, Howes, and Reductions from Fulton, L., Howes, T. and Hardy, J. (2004Biofuel for transport: An International Perspective, Paris, France, International Energy Agency

73. Newest vehicles in Greene et al2004Growing Energy; How Biofuels Can Help End America's Oil Dependence, Washington, DC, Natural Resources Defence Council, December

2
CHAPTER

BIOFUELS AND THEIR CO-PRODUCTS AS LIVESTOCK FEED: GLOBAL ECONOMIC AND ENVIRONMENTAL IMPLICATIONS

József Popp [1,†], Mónika Harangi-Rákos [1,†], Zoltán Gabnai [2,†], Péter Balogh [1,†], Gabriella Antal [1,*,†] and Attila Bai [2,†]

[1] *Institute of Sectoral Economics and Methodology, Faculty of Economics and Business, University of Debrecen, Debrecen 4032, Hungary*

[2] *Institute of Business Economics, Faculty of Economics and Business, University of Debrecen, Debrecen 4032, Hungary*

ABSTRACT

This review studies biofuel expansion in terms of competition between conventional and advanced biofuels based on bioenergy potential. Production of advanced biofuels is generally more expensive than current biofuels because products are not yet cost competitive. What is overlooked in the discussion about biofuel is the contribution the industry makes to the global animal feed supply and land use for cultivation of feedstocks. The global ethanol industry produces 44 million metric tonnes of high-quality feed, however, the co-products of biodiesel production have a moderate impact on the feed market contributing to just 8–9 million tonnes of protein meal output a year. By economically displacing traditional feed ingredients co-products from biofuel production are an important and valuable component of the biofuels sector and the global feed market. The return of co-products to the feed market has agricultural land use (and GHG emissions) implications as well. The use of co-products generated from grains and oilseeds can reduce net land use by 11% to 40%. The proportion of global cropland used for biofuels is currently some 2% (30–35 million hectares). By adding co-products substituted for grains and oilseeds the land required for cultivation of feedstocks declines to 1.5% of the global crop area.

Keywords: bioenergy; biofuels; co-products; feed potential; substitution

1. INTRODUCTION

In the last 35 years global energy supplies have nearly doubled but the relative contribution from renewables has increased from 13% to 19%, including about 9.3% from traditional biomass and about 9.7% from modern renewables (Figure 1). The contribution of "modern" renewables (e.g., solar, wind, biofuel) is still a marginal component of total global renewable energy supply, however, they are continuously growing. A major impetus for the development of bioenergy has been the search for alternatives to fossil fuels, particularly those used in transportation [1].

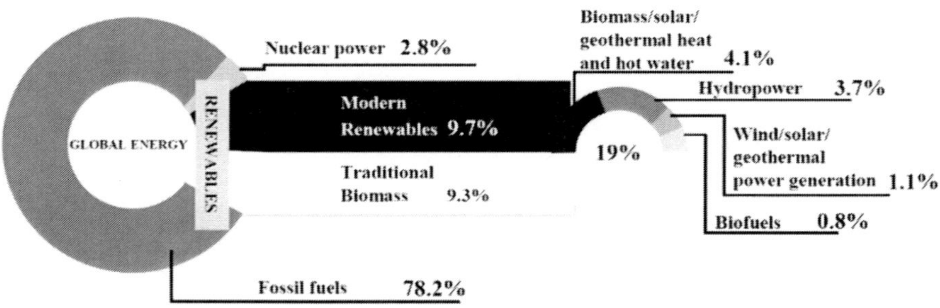

Figure 1. Estimated renewable energy share of global final energy consumption in 2011. Source: [2].

The transport sector is responsible for about 20% of world total energy use. Transport biofuels are currently the fastest growing bioenergy sectors, even though they represent around 3%–4% of total road transport fuel and only 5% of total bioenergy consumption today. Most capacity expansion and financing need is expected for next generation biofuels in the longer term and strong competition from other renewable energy projects with lower risks (wind and solar) can be experienced. Liquid biofuels for transport are generating the most attention, although just a small fraction of biomass is used globally for biofuels production at present [1].

Many countries support production and use of liquid biofuels for transportation to enhance domestic energy security, spur economic development and reduce emissions of greenhouse gases (GHG) and other pollutants. There is an increased focus on advanced biofuels in the past few years. The actual production volumes were not available, the installed production capacity values shows that the global advanced biofuel production capacity increased from 1.58 billion litres in the year 2010 to 4.21 billion litres in the year 2013 [3]. While considerable research and development is under way to commercialize new types of biofuel and feedstocks, the two primary biofuels produced globally today—ethanol and biodiesel—are predominantly derived from agricultural commodities, such as grain, sugar and oilseeds.

Currently, around 80% of the global production of liquid biofuels is in the form of ethanol. In 2012–2014 on average global fuel ethanol production reached 108 billion litres and global biodiesel production amounted to 28 billion litres (Figure 2 and Figure 3). The two world's top ethanol producers, the United States and Brazil, accounted for around 75% of total production. Biodiesel production is far less concentrated than ethanol. The European Union remained the centre of global biodiesel production, with 12 billion litres representing 43% of total output. Global expansion of biofuel production is projected to continue during the next decade, although at a slower pace than over the last half decade. Ethanol production in the United States is projected to be relatively flat over the next decade due to the ethanol blend wall and declining gasoline use. Most of the additional ethanol production is expected to take place in Brazil [4]. Indonesia will surpass the United States and Brazil in the latter years of the outlook period to become the second largest biodiesel producer behind the European Union. Global ethanol and biodiesel production are both expected to expand to reach, respectively, almost 135 and 39 billion litres by 2024 [5]. In 2040 the share of biofuels in road transport fuels would range—depending on policies—from 5% to 18% globally, from 11% to 31% in the European Union and from 11% to 29% in the United States [6].

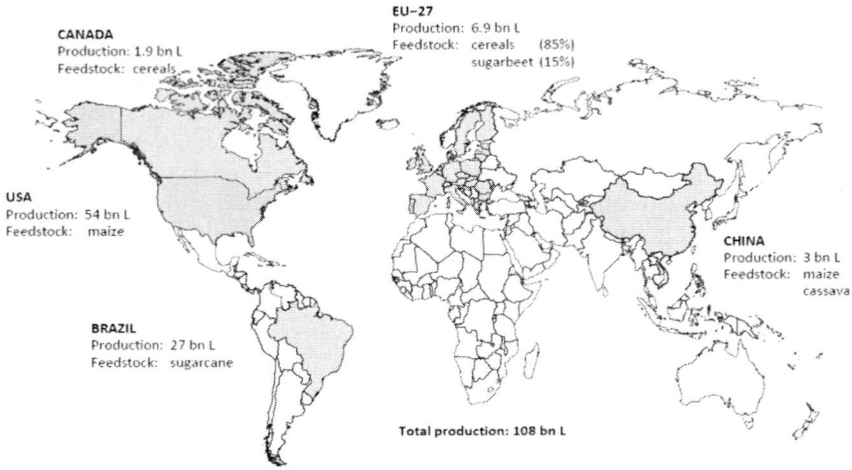

Figure 2. Word fuel ethanol production, average 2012–2014 [5].

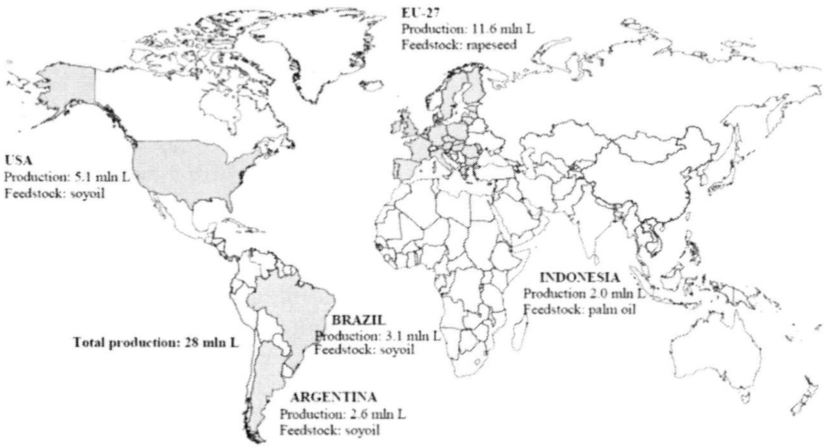

Figure 3. Word biodiesel production, average 2012–2014 [5].

The share of bioethanol in total grains demand—i.e., corn, wheat and other coarse grains—is about 6% of global production. The bulk of the worldwide use of grains in alcohol production comprises maize in the USA and China. However, an increase in the offtake of wheat for fuel ethanol can also be observed in Canada and the European Union. The share of biodiesel in rapeseed, soybean and palm oil demand is around 11% of global vegetable oil production. The share of waste biodiesel feedstocks such as animal fat and used cooking oil increased to 15% in total biodiesel output [7].

Increased biofuel production has led to criticism and concerns about food availability while it is feared that rising demand for agricultural land will lead to deforestation, grassland conversion and increased Greenhouse Gas (GHG) emissions from these land use changes. The main criticism is based on expected impacts of biofuel production following the introduction of dedicated biofuel targets and policies [8,9]. Most models were developed to evaluate agriculture or climate policies and were later adapted to incorporate biofuel production [9,10]. Early applications did not consider generation of co-products (co/by-products of the biofuel production process which are mostly used as animal feed) while second-generation biofuel production technology, at least in early applications, was not included [10,11]. Other restrictions include limited ability to adjust to accelerations in yield improvement or to changes in crop rotation [12]. Central to the debate on the impact of biofuel production is the way increased biomass requirements are to be met by area expansion, yield improvement or by increased cropping intensity. While the exact consequences of these limitations remain unclear economic model impact assessments of biofuel policies should be considered with care.

Following changes in biofuel policies a strong expansion in biofuel production was observed in the USA, the EU, China, and many other countries. These increases, however, were not sufficient to fully satisfy biofuel policy objectives in the USA and the EU. Further expansion of biofuel production is expected in the USA, Brazil, Argentina, and the EU. Co-product generation in early biofuel impact assessments was ignored leading to an overestimation of land requirements and GHG emissions. The output of feed co-products is relatively high in the USA, the EU and China due to the large share of grains used in ethanol production with high feed yields. It is low in Brazil where ethanol production is dominated by sugarcane which generates no feed co-products. Co-product yields are low for rapeseed and soybean used in the biodiesel industry. Estimates on impacts of biofuel production often use models with limited ability to incorporate economic and environmental implications by ignoring generation of co-products from biofuel production. Significant research and development efforts are under way to commercialize "second generation" feedstocks and biofuels, however, these are unlikely to be produced in quantity in the short term. Furthermore, the co-products from many of these new feedstocks are not likely to have applications in the animal feed market [4,13].

The review is organized as follows: first, it describes bioenergy potential including competition between conventional and advanced biofuels. Next, it presents feedstocks and co-products of biofuel production. Furthermore,

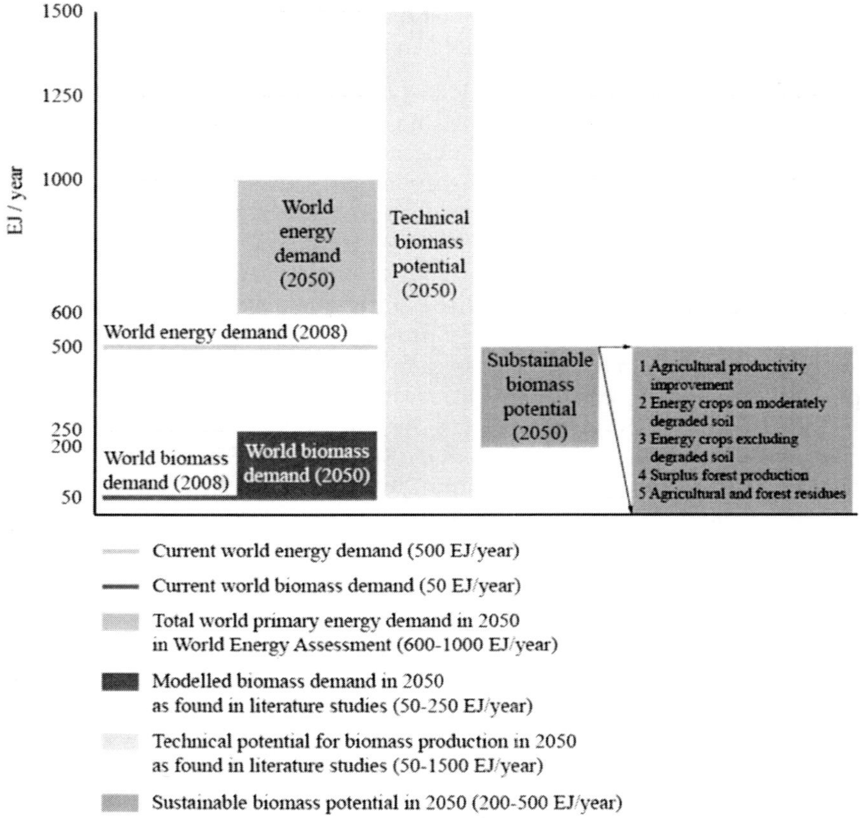

Figure 4. Global bioenergy sources [14].

attributes of co-products are carefully analysed based on economic and environmental implications of biofuel production. This is followed by materials and methods, and some conclusions.

2. RESULTS AND DISCUSSION

2.1. Bioenergy Potential

The worldwide potential of bioenergy is limited because all land is multifunctional and land is also needed for food, feed, timber and fibre production, as well as for nature conservation and climate protection. In addition, the use of biomass as an industrial feedstock (e.g., plastics) will become increasingly important. At present, some 55 EJ/year of bioenergy are produced globally. Modern forms of bioenergy in use in 2011 amounted to 23.6 EJ as heat, biofuel and electricity. An additional 31.4 EJ of traditional biomass was used very inefficiently for cooking/heating in poor rural areas, mainly in Africa [2]. Based on this diverse range of feedstocks, the technical potential for biomass is estimated in the literature to be possibly as high as 1500 EJ/year by 2050 (Table 1). However, most biomass supply scenarios that take into account sustainability constraints, indicate an annual potential of between 200 and 500 EJ/year (excluding aquatic biomass owing to its early state of development), representing 40 to 100% of the current global energy use [14]. Forestry and agricultural residues and other organic wastes (including municipal solid waste) would provide between 50 and 150 EJ/year, while the remainder would come from energy crops, surplus forest growth, and increased agricultural productivity (Figure 4).

Table 1. Statistical estimates of minimum and maximum values of global bioenergy potential (EJ/year).

Studies Referring to 2050	Low Range	High Range
Smeets *et al.* [15]	215	1272
IEA Bioenergy [16]	50	1500
Dornburg *et al.* [17]	200	500
IPPC [18]	50	500
Haberl *et al.* [19]	160	270
Global Energy Assessment [20]	80	140

Projected world primary energy demand by 2050 is expected to be in the range of 600 to 1000 EJ/year compared to about 500 EJ in 2008. The expert assessment suggests potential deployment levels of bioenergy by 2050 in the range of 100–300 EJ/year. However, there are large uncertainties in this potential, such as market and policy conditions, and there is strong dependence on the rate of improvements in the agricultural sector for food, fodder and fibre production and forest products. The entire current global biomass harvest would be required to achieve a 200 EJ/year deployment level of bioenergy by 2050. Scenarios looking at the penetration of different low carbon energy sources indicate that future demand for bioenergy could be up to 250 EJ/year [21] It is reasonable to assume that biomass could sustainably contribute between a quarter and a third of the future global energy mix (Figure 4). The total annual aboveground net primary production (the net amount of carbon assimilated in a time period by vegetation) on the Earth's terrestrial surface is estimated to be about 35 Gt carbon, or 1260 EJ/year assuming an average carbon content of 50% and 18 GJ/t average heating value [22], which can be compared to the current world primary energy supply of about 500 EJ/year [14]. All harvested biomass used for food, fodder, fibre and forest products, when expressed in equivalent heat content, equals 219 EJ/year [23]. The global harvest of major crops (cereals, oil crops, sugar crops, roots, tubers and pulses) corresponds to about 60 EJ/year and the global industrial roundwood production corresponds to 15 to 20 EJ/year [24].

Availability of land for non-food crops will be determined by increased yield potential, reducing losses and wastes along the food chain and lower inputs. However, these volumes will remain limited relative to total energy and transport sector fuel demand. Limited biomass resources will be allocated to the sector (materials, chemicals, energy) that is most able to afford them. This will depend on the price of existing fossil fuel products and the relative cost of converting biomass into substitute final fuels such as bio-derived electricity, ethanol blends, biodiesel and bio-derived jet fuel. It will also depend on factors such as cost of alternative fuel and energy sources, government policies including excise rates, and the emission intensity of each sector. The sustainable use of residues and wastes for bioenergy, which do not require any new agricultural land and present limited or zero environmental risks, needs to be encouraged and promoted globally. Several factors may discourage the use of these "lower-risk" resources. Using residues and surplus forest growth, and establishing energy crop plantations on currently unused land, may prove more expensive than creating large-scale energy plantations on arable land. In the case of residues, opportunity costs can occur, and the scattered distribution of residues may render it difficult in some places to recover them [25]. Whatever is actually realised will depend on the cost competitiveness of bioenergy and on future policy frameworks, such as greenhouse gas emission reduction targets. The uptake of biomass depends on biomass production costs, logistics, and resource and environmental issues [18].

2.2. Competition between Conventional and Advanced Biofuels

Significant development of advanced biofuels is necessary for diversification and decarbonisation of transport in the longer term. Although a number of projects on 2nd generation bioethanol ended with the opening of pilot and demonstration plants around the world and in spite of several proclamations, none of them is operating at the industrial scale. To make this possible, further reductions in processing costs will be necessary to achieve a product that is competitive with 1st generation bioethanol. By-products, e.g., lignin separated after pre-treatment procedure can be used to generate energy for ethanol plant operations or used as a dispersant and binder in concrete admixtures, as an alternative to phenolic and epoxy resins, or as the principal component in thermoplastic blends, polyurethane foams or surfactants. A combination of 1st and 2nd generation feedstocks (e.g., corn cobs together with stover) can eliminate bottlenecks and lead to product competitiveness [26].

Biofuels from dedicated lignocelluloses energy crops on marginal land is likely to be a cost-efficient contribution. However, extreme territorial and climatic conditions resistant species and varieties are required. Perennial, herbaceous crops—such as Miscanthus sp., A. donax—may be sustainable because the annual soil cultivation increases the air's carbon-dioxide level and these plants can mobilize mineral nutrients from the stems and leaves to rhizomes at the end of growing season, reducing the fertilizer needs. These species can rehabilitate the quality of marginal land [27,28,29,30]. Further requirements for energy plants that growth rapidly produce high biomass without irrigation (high water-use efficiency) and resistance to pests and diseases [31]. On the other hand, some potential biomass plants, such as giant reed, have invasive potential widely in riparian areas. Particularly, during the floods along the rivers, fragments of shoot and rhizome of giant reed can disperse and populate quickly and easily new habitats [32,33,34].

Since 2013 advanced biofuels have made good progress, with nine commercial-scale plants commissioned, seven of these in 2014 and 2015. Policies that mandate blending levels and provide capital incentives, along with the development of secure local feedstock supply chains, have been fundamental. New projects may require oil prices around USD 100/bbl or above to be attractive. There is significant potential to reduce the costs but a sustained long-term policy commitment would be needed, which may face risks from the lower oil price environment [35].

IEA [25] refer to the higher water consumption in the production of advanced compared with conventional biofuels, referring specifically to lignocellulosic ethanol. However, the major part of water consumed would be for feedstock cultivation, so that biofuels from wastes and residues should have a reduced water footprint. Local impacts on water quality and availability should nevertheless be monitored [36]. The processing of feedstocks into biofuels via advanced biochemical or thermochemical conversion routes can require relatively high energy inputs. However, other environmental impacts resulting from the processing of biomass through advanced conversion technologies such as water consumption in processing should be investigated and if necessary be addressed by safeguards [35].

Even as several conventional biofuel production facilities closed their doors, several advanced biofuel production facilities came on line in 2014. These included three new biorefineries using cellulosic plant material (predominantly corn stover) in the United States: POET-DSM, DuPont, and Abengoa. In Brazil, three commercial, 2nd generation biofuel projects started operation: GranBio commercial cellulosic ethanol plant, Raizen/Iogens plant, and Solazyme-Bunge plant. The advanced biofuels industry also faced challenges, however. US-based Kior filed for bankruptcy and decommissioned its commercial-scale cellulosic biofuel plant in Mississippi. In Europe, EU policy requires that every Member State obtain 10% of its transport fuel from renewable sources by 2020, however, new legislation limits the contribution of biofuels derived from sugars, starch, and oil crops due to sustainability concerns, which are mainly about indirect land-use change. This, in combination with amendments to some national biofuel policies, has raised uncertainty among producers. So development of advanced biofuels has lagged due to the lack of EU-wide policy support, although some Member States have started to enact national policies, for example Italy announced a mandate of 0.6% advanced biofuels by 2018 [37].

Cellulosic biofuels may use crop residues or other wastes, but most plans for these biofuels rely on planting and harvesting fast-growing trees or grasses. At least some direct competition with food is still likely because such trees and grasses grow best and are most easily harvested on relatively fertile lands already dedicated to crops. Using cropland to grow trees and grasses rather than food crops for biofuels will probably not reduce, let alone eliminate, competition for cropland. Trees and grasses will have a hard time producing more biofuels per hectare than today's crop-based biofuels. Alternatively, cellulosic biofuels might rely on harvesting existing forests or producing fast-growing trees or grasses on the world's grasslands or woody savannas. Some researchers also point to abandoned farmland as a candidate for bioenergy production that avoids competition for land. But abandoned farmlands typically regenerate into forests, woodlands, or grasslands if left alone, which provide climate benefits that are already assumed and counted in climate change assessments. These benefits would be sacrificed by using that land for bioenergy. Although there is capacity to increase plant production on each hectare by increasing yields or by enhancing use of lands that are degraded, however, that capacity is already needed to meet rising demands for food and wood products while preserving ecosystems and their carbon. There are some biomass feedstocks that avoid the competition for land, namely various forms of wastes and residues. In the long run, such wastes might contribute modestly toward replacing some uses of fossil fuels. Phasing out the dedicated use of land to generate bioenergy, particularly biofuels, would reduce the food gap and, perhaps even more importantly, keep it from greatly expanding [38].

Searchinger and Heimlich [38] defined the dedicated use of land for bioenergy as the production of bioenergy that sacrifices alternative outputs from land. This narrow definition treats bioenergy production in isolation. For instance, some feedstocks (such as soybeans and rapeseed) for biofuel production would not fit into this definition because the land produces protein as animal meal as the main product and, secondarily, oil (either as cooking oil or a biofuel feedstock as biofuel products). That is, biofuel production from oil would not cause dedicated use of land for bioenergy production. The authors certainly did not promote bioenergy production from those feedstocks. Also, "land" can be very different in terms of suitability and productivity for different vegetation types. Land that may not be suitable for row crops could be suitable for growth of other vegetation types because of differences in nutrients, water, climate, and other requirements. Further, the authors treated marginal/degraded lands as not existing [39].

Searchinger and Heimlich [38] correctly pointed out that directing wastes like crop residues and municipal solid waste to biofuel production is a good use of these resources. They raised concerns, however, that dedicated cellulosic crops are not a promising option for a biofuel feedstock because they require land and do not have sufficiently high yield. Some of the studies on this topic indeed indicate that doubling of currently low cellulosic biomass yields is achievable [40]. Reliable reports indicate that significant areas of marginal lands exist that could be used to produce cellulosic crops that are currently underutilized [41]. Many studies have identified land where cellulosic biomass can be grown to avoid competition between cellulosic biomass and food production [42,43].

Conventional biofuels are currently produced in many countries and are based on well-known processes and feedstock. Apart from sugarcane ethanol, conventional biofuels will hardly be sustainable in the future as large-scale production would take away feedstock and land from food production and forestry. In addition, they are rather expensive and offer only limited reductions in greenhouse gas (GHG) emissions compared to fossil fuels.

Advanced biofuels promise to be more sustainable, with higher emissions reductions. They are based on biomass resources and land not used for other primary needs, such as food production and farming. Feedstock includes lignocellulosic residues from agriculture and forestry, fast-rotation non-food crops (possibly grown on marginal, non-arable land), organic fraction of urban waste and micro-algae. The conversion of these resources into biofuels requires processes that are currently under commercial demonstration or under development, with small plants in operation and large plants under construction or planned all over the world.

2.2.1. Giant Reed for Biofuels

Non-food energy crops, such as woody plants (poplar, willow, Eucalyptus sp.) or herbaceous (Miscanthus sp., switchgrass or giant reed), is produced for bioenergy, biogas and biofuel purposes [44,45]. Arundo donax L. (common name "giant cane" or "giant reed") is a perennial, rhizomatous species which has been introduced around the world by humans as an ornamental/crop plant. Giant reed is a sterile plant without any viable seeds, it can be propagated vegetatively from the rhizome or stems [10,35]. Propagules can be produced also by hydroponic or in vitro micropropagation methods [36]. Among the promising biomass plants, A. donax has high biomass yield per hectare and can be adapted to different types of soil and water condition [31,46]. It has lower agronomic input requirement than traditional crops [47]. It can be a good candidate to restore and recover soil ecosystem [48,49] and has ability to remediate the contaminated soil and water [50,51]. There is increasing commercial demand for giant reed production [9,10], however, there are limited data available about the production of biogas or bio-ethanol from giant reed.

From the second half of the 1990s, giant reed is regarded as one of promising plants of the biomass industry due to high biomass production per hectare [10,37]. For example in Central Italy, a 12-year field trial without irrigation could produce 38 tonnes dry matter per hectare per average year [38] and 20 tonnes dry matter with no fertilization on sandy soil [39]. For its cultivation low agronomic and energetic input is required [10,40]. Due to the high biomass per hectare and chemical composition of giant reed (Table 2.) large amounts bio-ethanol can be produced. Williams et al. [49] and Jaradat [50] reported same bio-ethanol production (11,000 $L \cdot ha^{-1}$) in case of 45 tones ha^{-1} biomass yield. According to Corno et al. [14] 12,960–15,228 $L \cdot ha^{-1}$ bio-ethanol can be produced, which is higher than reported from other energy crops for example than Miscanthus x giganteus [31].

Table 2. Chemical composition of A. donax.

Hemi-Cellulose (%)	Cellulose (%)	Lignin (%)	Ashes (%)	References
24.2	41.6	24.9	3.2	[52]
34.8	20.9	23.0	n.d.	[53]
25.61 ± 0.07	33.85 ± 0.06	24.02 ± 0.04	5.04 ± 0.03	[54]
24.4 ± 0.52	39.1 ± 0.25	19.2 ± 3.25	4.2 ± 0.67	[55]
14.5	39.6	24.3	5.3	[56]
35.27 ± 2.80	31.10 ± 1.03	18.49 ± 0.10	n.d.	[57]

Source: Respective authors' data.

Despite increasing interest in giant reed, propagation and production systems should be optimised reducing costs of bioenergy and bioproducts production to enhance economic feasibility. For biofuels production, it is necessary to improve sugar yield and reduce lignin content, therefore there have been more studies concerning various pretreatments of giant reed. Energy inputs and efficiency of converting giant reed to bioenergy and bioproducts are significantly influenced by the composition characteristics, such as minerals, cellulose, hemicelluloses, lignin, moisture content and recalcitrant compounds. It can be cultivated almost in all climatic zones. Cold seems to be a limiting factor, therefore researchers have started to develop cold-resistant giant reed varieties [36,41]. It has invasive potential in Mediterranean or warmer climates, so in case of establishment of giant reed plantations surrounding water conservation areas have to be avoided to reduce potential hazards.

2.2.2. Algae for Biofuels

Algae have been cultivated commercially since the 1950s, mainly for the pharmaceutical industry, but have only recently gained attention as a potential source of biomass. Third-generation biofuel usually means biofuel from algae. To date, there have been numerous studies of algae and other water based biomass in order to identify strong candidates for biomass accumulation rates as well as lipid content for production of biodiesel. Marine and aquatic biomass can be a useful alternative source of biomass that can be used to produce a wide range of biofuels for commercial use. Primarily, growth of algae for the production of oils and energy conversion has focused on microalgae. Algae have potential as a feedstock for biofuels. Depending on their composition, different algae species may be suitable for a range of biofuels. Algae promise a potentially high productivity per hectare, could be grown on non-arable land, can utilise a wide variety of water sources (fresh, brackish, saline and wastewater), and potentially recycle CO_2 and other nutrient waste streams. However, algae cultivation faces several challenges, related to availability of locations with sufficient sunshine and water, required nutrient inputs, and oil extraction [58].

Algae feedstocks for alternative fuels production are not economically competitive with fossil fuels at the present time. Furthermore, it has not yet been demonstrated that algae production systems offer improved sustainability characteristics. Like fossil fuel, algae fuel releases CO_2 when burnt, but unlike fossil fuel, algae fuel and other biofuels only release CO_2 recently removed from the atmosphere via photosynthesis as the algae or plant grew. Additionally, algal biomass productivity per hectare could eventually be higher than for terrestrial energy crops. Based on the high level of innovation demonstrated within the algal biofuels industry in just the past decade, it is likely that new technologies will continue to be introduced in the future. In fact, the introduction of these innovations will be critical if the sector is ultimately going to achieve commercial success [59,60].

Growth of aquatic and marine biomass is not without challenges because maximum growth rates of the microorganisms typically occur under very specific conditions. Furthermore, open pond algal systems are susceptible to contamination from various airborne microorganisms that can decrease overall productivity. But the main concern is the ability to separate algae from water, which due to their very dilute nature, can be expensive and inefficient. However, it is feasible to use algae as a means for tertiary wastewater treatment in order to utilize trace nutrients such as phosphorous- and nitrogen-containing compounds, or can be used at industrial processes as a way to absorb carbon dioxide by entraining algal cultures to gaseous exhaust streams [61].

Photosynthetic microorganisms such as microalgae and cyanobacteria could serve as an attractive feedstocks because they have higher growth rates requiring much less land area compared to plants and they can thrive in areas that cannot usually support mainstream agriculture. Algae can be cultivated at sea or on non-arable land, so there is no competition with current food production [62]. However, despite significant progress, reliable and cost-effective production of lipid- and protein-rich algal biomass have not been demonstrated at scales $> 10\ m^2$. Productivity and cost remain the two fundamental barriers to commercialization [63]. Commercially viable production of biofuel from algae will depend on effective strategies to generate high-volume, low-value biofuel along with high-value co-products.

Several companies and government agencies are funding efforts to reduce capital and operating costs and make algae fuel production commercially viable. Commercial cellulosic biofuel production began in the US in 2013, while algae biofuels are not yet produced commercially [6,37].

Recently, microalgae have emerged as a source than can play the dual role of bioremediation of wastewater and generation of biomass for biodiesel production. Growing microalgae on different types of wastewaters has been studied over the past decades. The success of such studies depends on the performance of the selected microalgae strains [64]. Moreover, the use of wastewater treatment becomes an economically attractive alternative [65]. Various types of algal biofilm reactors integrated with wastewater treatment were developed to overcome current limitation of algal biofuel production. So far the integration of algal bioreactor and wastewater has been limited to municipal wastewater while only a few agricultural wastewater have been used for algal bioreactors. However, algal biofilm reactors integrated with wastewater would have high potential for high productivity of algal biomass for biofuel and efficient wastewater treatment if various conditions are optimized [66]. In addition, a prospective life cycle assessments (LCA) of two algal systems were studied. The LCA impact results conclude that algal system producing biodiesel, animal feed, and succinic acid could be beneficial to the environment compared to that algae system that produces biodiesel and animal feed. Thus algae system could be a potential alternative or additional renewable system to mitigate environmental impact [67].

2.3. Feedstocks and Co-Products of Biofuel Production

Sugarcane is the predominant feedstock for ethanol production in tropical regions (Brazil). In temperate areas, ethanol is mostly made from cereals (maize in the USA and China, wheat in the EU and China). Main biodiesel feedstocks are soybean (Brazil, USA), rapeseed (EU), and oil palm (Indonesia and Malaysia). There are other feedstocks of minor importance, such as castor beans in Brazil, sunflower in the EU and jatropha in Mozambique, but these are not included in the analysis. Crop yield is high for sugarcane (Brazil, South Africa), sugarbeet, and oil palm. Cereal yields are high for corn in the USA, but less so for corn and wheat in the EU and China. Rapeseed and soybean yields are modest. Ethanol yields are highest for sugarbeet, and sugarcane (Brazil). Highest biodiesel yields were observed for oil palm (Indonesia, Malaysia). Of the four largest sources of biofuels: maize, sugarcane (bioethanol), soybean and rapeseed (biodiesel), only sugarcane appears to have a secure future although in the case of maize this will depend on the rate of yield improvement that could be achieved. The progressive divide between increase in demand and increase in production suggests that maize ethanol will lose long-term economic viability, unless the already high rate of yield per hectare improvement can be accelerated yet further. With the exception of oil palm, the yields of biodiesel crops (soybean and rape/canola) are too low to contribute significantly to future energy supply. Soybean and rapeseed produce far too little fuel per unit land area to remain competitive without mandates and subsidies. However, breakthroughs in engineering accumulation of oil in vegetative tissues may provide an alternative with the potential of developing a sugarcane that accumulates oil in place of sugar [68].

The bioethanol share in total grains demand—i.e., corn, wheat and other coarse grains—is about 6% of global production. The fuel ethanol sector accounts for 13% of global maize consumption and 20% of global sugar cane production. An estimated 143 million tonnes of grain is used globally for ethanol. The US is the global leader in grain ethanol production, accounting for roughly 90% of total grain use for ethanol, followed by the European Union, China and Canada. Maize accounted for the majority of grain use for ethanol in the United States and China. European and Canadian producers principally use wheat and maize for ethanol. About one-third of the volume of grain processed for ethanol was used to produce animal feed as co-products, thus, the equivalent of two-thirds of the of grain were used to produce fuel. Feed market impacts of increased maize use for ethanol are smaller than that indicated by the total amount of maize used for ethanol production because of DDGS. While ethanol expansion raised demand for maize, DDGS partially offsets the impact on the feed market. Consequently, the net effect in the domestic feed market of a tonne of maize being used for ethanol production is less than a tonne. For example, the amount of feed (corn and soybean meal) replaced by the DDGS represents about 38% (weight basis) of the maize used in the associated ethanol production process for a given crop year. Furthermore, grain use for ethanol is expected to moderate in accordance with slowing national mandates in the US and the EU [4].

The biodiesel share of rapeseed, soybean and palm oil demand is around 11% of global vegetable oil production. The share of waste biodiesel feedstocks such as animal fat and used cooking oil increased to 15% in total biodiesel output. Continuously growing demand for protein meal has been the main driver behind the expansion of oilseed production in recent years. This has increased the share of protein meal in the value of oilseeds and favoured soybeans over other oilseeds. Compared with coarse grains and other feed ingredients, protein meal prices have stayed relatively high. Global oilseeds production reached 530 million tonnes (soybean 320 and rapeseed 70 million tonnes) in the 2014/15 marketing year. At the same time soybean production increased faster than production of rapeseed, sunflower, increasing the sector's concentration. Vegetable oil production increased to 180 million tonnes (out of this 60 million tonnes palm oil). Demand growth has slowed in recent times due to stagnating biodiesel production from vegetable oils in developed countries. Production of rapeseed in Canada and the European Union is expected to grow much slower than in the previous decade as high oil-containing oilseeds like rapeseed are more affected by the slower growth in vegetable oil prices [7]. Vegetable oil use in biofuel production account for about 20 million tonnes a year. Soybean and rapeseed oil has a 70% share of the total feedstocks used in biodiesel production worldwide. Oilseeds such as rapeseed or canola and soybeans represent the most common source of vegetable oil feedstocks for biodiesel production. An estimated 6 million tonnes of rapeseed oil and 7 million tonnes of soybean oil is used globally in the production of biodiesel, representing roughly 70% of the total feedstocks used in global biodiesel production [6,7].

Biofuel production from wheat, maize, rapeseed and soybean yields valuable protein-rich co-products such as rape meal, soymeal and dried distillers' grains and solubles, which can be used as animal feed in the livestock industry. Because protein-rich crops generally require a relatively large amount of land for a given output compared with cereal crops, the use of co-products can reduce net land use. Co-products considered in this study include dried distillers' grains with solubles (DDGS), soy meal, rapeseed meal. Palm kernel cake and sugar beet pulp is an important, though low quality component of feed concentrates. These co-products are produced in insignificant amounts and are therefore not considered in this analysis.

Animal Feed Produced from the Ethanol and Biodiesel Industry

Over 90% of the ethanol produced today comes from the dry milling process and 10% from wet milling. Both the wet and dry mill processes utilize only the starch portion of the corn kernel for ethanol production. The remaining protein, fat, fiber and other nutritional components remain available for use as animal feed. In distillers dried grains with solubles (DDGS) or distillers dried grains (DDG), these remaining nutritional components from the corn kernel are essentially concentrated by a factor of three, meaning typical distillers grains have at least three times as much protein and fat as an equivalent amount of corn. If the distillers grains are being fed to livestock in close proximity to the ethanol plant, the drying step can be avoided and the product is called wet distillers grains (WDG). An estimated 85% of existing dry mills have the capability to extract corn oil, which is then sold as an individual feed ingredient or as a feedstock for biodiesel production. In the wet milling process corn oil from the germ is either extracted on-site or sold to crushers who extract the corn oil. The remaining fiber, gluten and starch components are further segregated and sold as corn gluten feed (CGF) or corn gluten meal (CGM). The remaining starch can then be processed in one of three ways: fermented into ethanol, dried or modified corn starch, or processed into corn syrup.

In fact, one-third of every bushel of grain that enters the ethanol process is enhanced and returned to the animal feed market, most often in the form of distillers grains, corn gluten feed and corn gluten meal. These nutrient-dense co-products are fed to beef cattle, dairy cows, swine, poultry, and fish in nations around the world. In the 2013/14 marketing year, the U.S. ethanol industry produced an estimated 39 million metric tonnes, the EU 4 and China 2 million metric tonnes of high-quality feed, namely 41 million tonnes of distillers grains and around 4 million tonnes of gluten feed and gluten meal. Estimated average potential U.S. DDGS feed consumption is around 62 million tonnes a year, an amount much higher than the current DDGS supply [16,18]. Corn gluten feed and meal production has remained relatively constant in the last decade, as the majority of growth in ethanol production has come from dry mills. As for the wet milling co-products, corn gluten feed is primarily fed to dairy and beef cattle, while corn gluten meal is used to feed a wide variety of species, including poultry and fish. Corn gluten meal, which features a very high concentration of protein, is often used as an ingredient in pet food as well. While much has been learned over the past decade about the benefits of feeding co-products, much more research is under way at public and private institutions to increase co-product inclusion levels to best take advantage of their nutritional and economic benefits. While co-products of the ethanol industry are used by the domestic markets in the EU and China the United States exported around 10 million tonnes of distillers grains and approximately 2 million tonnes of corn gluten feed and corn gluten meal—roughly 28% of total U.S. production—in to nearly 50 countries worldwide in 2013. These exports in 2013 were volumetrically equivalent to about 24% of total 2013/14 corn grain exports. Increases in ethanol co-product exports are supplementing corn export levels, meaning total exports of corn and corn products continue to trend upward [69,70].

By economically displacing traditional feed ingredients, ethanol co-products effectively reduce the livestock and poultry industry's demand for maize and protein meal. The return of co-products to the feed market has agricultural land use implications as well; at least one-third of every hectare "dedicated" to ethanol production should actually be characterized as producing feed, not fuel. Co-products from grain ethanol production are an increasingly important and valuable component of the biofuels sector and the global feed market. Today, more than 85% of existing dry mills have the capability to extract corn distillers oil (CDO) in the U.S., which is then sold as an individual feed ingredient or as a feedstock for biodiesel production. Distillers grains are increasingly recognized as an extremely efficient feed source. In many cases, co-products are a more effective source of energy and protein than the ingredients they are replacing in the diet. Distillers grains provide approximately 130%–150% of the energy of an equivalent amount of corn when fed to beef.

Historically, distillers grains have generally sold at a significant discount to competing feed ingredients, meaning they are a good value for livestock and poultry feeders around the world. Over the past several years, DDG prices have consistently sold for 75%–90% of the Central Illinois corn price and 40% of Central Illinois soybean meal prices. However, in 2011 and 2012, DDGS prices sold for 83% the price of corn in the Central Illinois reference market but just 47% the price of soybean meal. Moreover, as prices for soybean meal and other protein feeds spiked in 2013 due to the 2012 drought, DDGS was increasingly used as a replacement for protein meals. Thus, DDGS prices have averaged 109% the price of corn and 40% the price of soybean meal since the beginning of 2013 until mid-2014. A tonne of DDGS contains approximately 270 kilograms of protein, while a tonne of high-protein soybean meal contains 436 kilograms of protein. Thus, DDGS served as a lower-cost means of meeting protein needs during the period in which the 2012 drought affected global feed supplies [69].

Feed co-product output is expected to grow more slowly in the coming years. However, a number of new and emerging technologies may change the composition and further improve the quality of feed co-products. As discussed earlier, about 85% of dry mills today extract CDO from the stillage on the back end of the process. This CDO can be sold into the feed market (particularly for poultry) or used as a feedstock for biodiesel. When used as a biodiesel feedstock, CDO displaces higher value fats and vegetable oils that are typically used in food or feed applications. The feed co-products resulting from oil extraction practices typically have lower fat and higher protein content than conventional distillers grains. These "reduced fat" co-products can be more useful for certain species. Dry fractionation is another technology that may emerge more broadly in the dry mill ethanol industry. Essentially, this practice allows dry mill ethanol producers to separate the corn germ and other components from the starch on the front end of the ethanol process. The germ can then be processed and sold as feed or as feedstock for further processing for other uses. Further, some dry mill facilities are examining the potential of converting corn kernel fiber to cellulosic ethanol. This process would reduce the fiber content of DDGS, which may make the product more palatable at higher inclusion levels for monogastric animals. Additionally, a number of new and emerging ethanol processing aids are likely to improve the nutritional quality and utility of ethanol co-products. The success of these new technologies will depend on the demands of the ethanol industry's customers in the animal feed market. New technologies and practices promise to change the complexion of the ethanol co-products market in the years ahead.

The main components used to make biodiesel are soybean or rapeseed oil and an alcohol source, typically methanol. Seed meal co-products are left after oil is extracted from soybeans and canola oil. An estimated 80% of soybean seed 60% of rapeseed is left from the extraction process as seed meal, creating a significant quantity of this important co-product. The oilseed meal can be used immediately as an animal feed without further treatment. About 7 million tonnes of soybean oil and 6 million tonnes of rapeseed oil is used in biodiesel production contributing to almost 3.6 million tonnes of rapeseed meal and 5.6 million tonnes of soybean meal output [6]. Taking into consideration that 210 million tonnes of soymeal and 40 million tonnes of rapeseed meal is produced a year globally, the co-products of biodiesel production have a moderate impact on the feed market. The main direct by-product of biodiesel production is glycerine constituting about 10% of these materials on average. There is a limited demand for glycerin, the by-product of biodiesel production for a number of food, beverage, personal care and oral products, as well as pharmaceutical and other industrial uses. Glycerin can be used effectively in livestock rations to replace fossil-based glycerine. Importantly, glycerine can also be utilized as a feed ingredient for livestock rations. Crude glycerin contains similar energy to that of maize for pigs [24].

2.4. Attributes of Co-Products of Biofuels Production

The biofuel production process produces the fuel and other co-products. The type and quantity of co-products strongly depends on the biofuel production chain. The economic viability of the biofuel industry depends to a large extent on the ability of the industry to derive value from the biofuel it produces as well as the co/by-products that are generated during the process. When co-products are used credits can be attributed to the biofuel production chain. Credits include GHG emission savings, avoided land use or avoided energy use. The challenge is to decide how to distribute quantities of potential credits between the fuel and the co-products.

Credits for co-products are an important element in the calculation of greenhouse gas (GHG) reductions of the different biofuel production chains compared to fossil fuel use. For the evaluation of GHG emission savings common methods include mass based (allocation by mass), market value-based (allocation by price), and energy content-based (allocation by energy content) calculations. For reasons of feasibility the EC recommends in its regulatory framework of the proposal for a directive on renewable energy targets to apply allocation by energy content to determine GHG savings compared to fossil fuels [71]. However, for a fuel supplier it is not possible to know for certain what the substituted product really is and thus GHG emissions of these substituted products are uncertain. Moreover co-products used as animal feed can replace many different animal feed products, which have differing GHG performances.

Co-products can be applied as animal feed, substituting other feeds, such as maize and soymeal. When for example maize and soymeal is substituted, less land for feed cultivation is needed. This substitution reduces the indirect land use and therefore the impact of indirect land use change and intensification is substantially. The impact of the co-products on land use can be included. Recently several studies have attempted to include attribution of those co-products, which can be used as animal feed in an analysis of land use requirements of biofuels. When co-products are used for heat and process energy the energy balance improves. In those cases, feed has to be produced elsewhere and even ILUC effects cannot be excluded [1].

2.4.1. Economic Implications

Biofuels support policies in OECD countries are costly and there are alternatives to current support policies for biofuels that would more effectively allow governments to achieve their objectives. The impact of biofuels policies on GHG emissions is limited, however, biofuels support policies have significant impacts on global commodity prices. Alternative policy approaches may offer greater benefits including reduced energy demand and GHG emissions, and freer trade in biofuels. Further research can contribute to second generation biofuels from more efficient feedstocks, such as cellulose and other biomass [72]. Replacing fossil fuels with biofuels has the potential to generate a number of benefits. Biofuels can be produced domestically, which could lead to lower fossil fuel imports. Reducing the demand for petroleum could also reduce its price, generating economic benefits for domestic consumers, but also potentially increasing petroleum consumption abroad [73]. Bioenergy could enable farmers in poor regions to introduce agricultural technologies and improve infrastructure to increase farm productivity and thus to raise farmers' income [74]. Bioenergy and food production can co-exist and enhance each other by advancing technologies and increasing yields [75], thus complementing each other instead of competing against each other.

Feedstock generally accounts for around 70% of production costs, with processing, transportation, and other costs making up the remainder. Therefore, declining feedstock prices helped industry by reducing overall production costs. In 2014, most fuel ethanol was produced from sugar crops (roughly 61%), with the remainder from grains (roughly 39%). Feedstocks vary significantly depending on the country or region. For example, fuel ethanol production in the United States is based largely on corn, whereas Brazil relies primarily on sugar crops, and China on sweet sorghum, cassava, and other non-grain crops. Global biodiesel production is based largely on vegetable oils, mostly from rapeseed (Europe) and soybeans (United States, Brazil, Argentina), with smaller shares from palm (Indonesia) and other sources such as jatropha and coconut. Biodiesel production also includes industrial by-products such as used cooking oils (the main feedstock in China) and animal fat. In Europe, the relative share of cooking oil and tallow in biodiesel production is increasing as EU policy allows these feedstocks to be double-counted in transportation targets [37].

Increased acres for corn farming in the U.S, are driven by the significant increase in corn prices since 2005. However, several key factors cause increases in corn price, including U.S. corn ethanol production, weather events, recent grain demand increases, and diet changes in emerging economies [76]. Economic models show that biofuel use can result in higher crop prices, though the range of estimates in the literature is wide. Projections for the effect of biofuels on corn prices in 2015 ranging from a 5 to a 53 percent increase [77]. The possible impact of developed countries' biofuels policies on global food prices became a significant concern in 2007, when global grain prices reached historic heights. Though some experts associated the unprecedented price spikes in food grain and oilseed with these countries' biofuels policies [78,79,80], most of them now agree that these policies are unlikely to have been the main culprit, although they may have been a factor emphasizing that biofuel policy is only responsible for part of that fraction of price increases in food grain commodities that is due to biofuels [81].

Another study estimates that the impact of EU biofuels demand from 2000 until 2010 has increased world grain prices by about 1%–2% and oilseed prices by around 4%. It also estimates that without any cap on crop-based biofuels, EU policy could raise grain prices by 1%, and oilseed prices by 10% by 2020 [82]. Increasing the productivity of current and emerging bioenergy crops per unit land area is not only critical to economic viability, but also to biodiversity by minimizing the total land area needed. Land sparing is found far more effective than land sharing in strategies to realize bioenergy. Maize ethanol, often portrayed as the villain of the piece in the food versus fuel debate, may in fact have been key in stimulating yield improvement, including through genetically modified (GM) traits, that has resulted in increased exports of grain from the USA while providing a buffer in drought years [69].

Concerns over the role of food crops in the energy sector are well-founded, and for this reason (among others), the Renewable Fuel Standard caps corn ethanol volumes at 15 billion gallons. In short, for biofuels to expand their role in the transportation sector in the United States, they must be produced from cellulosic feedstocks, such as energy grasses, short rotation woody crops, crop residues, wastes, and other sources. In the EU new legislation limits the contribution of biofuels derived from sugars, starch, and oil crops due to sustainability concerns, which are mainly about indirect land-use change.

Biofuel co-products often substitute for higher priced feeds in animal rations. The increased use of agricultural commodities for biofuels has led to higher costs for animal feeds, however, increased substitution of co-products for traditional feedstuffs in feed rations mitigate input cost increases faced by livestock and poultry producers. Growth in the use of agricultural commodities for biofuels is expected to continue in the next 10 years, but with growth rates slowing in key producing countries as government-imposed limits on grain use for biofuels are reached and new non-agricultural feedstocks are commercialized [4,5].

DDGS and corn prices are highly correlated, and their correlation has strengthened in recent years. Soy and rapeseed meals have always been a major component of animal feeds, because they are excellent sources of protein. The impact of increased ethanol production is largely felt through competing crops as increased production of DDGS and oilseed meals can jointly reduce the demand of livestock industry for maize and oil meals and offset the increase in the demand of ethanol industry for grains due to the US and EU biofuel mandates. With increasing biofuel production the production of these co-products also increases. Prices of co-products are highly correlated with prices of feedstocks, such as grains and oilseeds and they represent an important component of total industry revenues. As a result, co-products prices fall relative to other feed ingredients. This encourages livestock producers to use more biofuel co-products in their production processes. On the other hand, any reduction in the prices of co-products diminishes total revenue and acts as a brake on growth of the biofuel industry. Biofuel co-products function as both a shock absorber and a price adjuster [83].

Between 1983 and 2006 the price of DDGS relative to maize has fallen by nearly 50%. This has provided a strong incentive for livestock producers to use more DDGS in their production process and has also enhanced US exports of DDGS [83]. The ratio of the average price of DDGS to the average price of maize reported for Iowa plants from 2007 through March 2015 ranged from 0.67 to 1.48 and averaged 0.91 for the entire period [84]. Further, the relative contribution of distillers' grains to gross returns has varied over time as the price of DDGS has varied.

The net contribution of DDGS to net returns for ethanol production is a function of the ratio of the price of DDGS to the price of maize. When the ratio is high, distillers grains provide a larger contribution to net returns and vice versa. With all other prices unchanged, a decline in the price of DDGS would result in a corresponding decline in net returns. For example, the current high ratio is contributing to net returns of ethanol production [84].

The global grain and oilseed supply has grown substantially in recent years and increased use of these commodities for biofuels production has not led to reduced availability for feed or food use. The amount of grain and oilseeds available for uses other than ethanol and biodiesel production is expected to grow more significantly in the long term, as feedstock use for biofuel production moderates in accordance with slowing national mandates.

2.4.2. Environmental Implications

Land Use Change and GHG Emission

Changes in land use, principally those associated with deforestation and expansion of agricultural production for food, contribute about 15% of global emissions of GHG. Currently, less than 3% of global agricultural land is used for cultivating biofuel crops and land use change associated with bioenergy represents only around 1% of the total emissions caused by land-use change globally most of which are produced by changes in land use for food and fodder production, or other reasons [85]. Indirect land-use changes, however, are more difficult to identify and model explicitly in GHG balances. Most current biofuel production systems have significant reductions in GHG emissions relative to the fossil fuels displaced, if no indirect LUC effects are considered.

Despite the availability of additional land, however, land use in agriculture has actually grown very slowly for decades. The best rainfed cropland is already being used, and expansion to other areas would incur higher input costs on average. Existing land is being used more intensively in most regions, as the practice of multiple cropping has spread, particularly in areas where land is relatively scarce, such as in Asia. Rising concerns and government regulations over climate change, biodiversity and resource management will push agriculture towards greater sustainability and lower environmental costs while competition between food and biofuels could also become more intense [86].

Changes in land use patterns may increase GHG emissions by releasing terrestrial carbon stocks to the atmosphere [87]. Biofuel feedstocks grown on land cleared from tropical forests, such as soybeans in the Amazon and oil palm in Southeast Asia, generate particularly high GHG emissions [88]. Even use of cellulosic feedstocks can spur higher crop prices that encourage the expansion of agriculture into undeveloped land, leading to GHG emissions and biodiversity losses [89].

Although biofuel production is only one of numerous contributors to land-use change, Fritsche and Wiegmann [90] estimate that by 2020 the overwhelming majority of land-use change—both direct and indirect—could be caused by feedstock production for biofuels. Because of its indirect nature, ILUC is a not a local phenomenon that can be observed in a given place and time. It cannot be monitored for individual feedstocks (e.g., biofuel crops), because production can be displaced anywhere in the world and because displacement can be distributed through global trading or occur with significant time lags.

Depending on the location of production, the total amount of area converted as well as the GHG emissions per hectare converted can vary greatly. And this global distribution of production depends on the assumptions made about the role of geography in international trade. Assumptions about yield response on the extensive and intensive margins are certainly critical in determining land-use change outcome and trade elasticity. Yield response for other commodities (other than the focus feedstock) and other regions are even more important than the yield response of the specific feedstock/region under examination (U.S. corn in this case). In addition to the intensive margin of yield response, changes in the assumption about crop yield response at the extensive margin can also have a dramatic impact on the resulting net global cropland requirement. Furthermore, consumer demand elasticity for food products play a critical role in determining the necessary expansion in global land-use following expansion of ethanol production [11].

The demand for land to produce bioenergy feedstock can cause shifts in land use patterns that can affect agricultural land. Land use change (LUC) impacts of biofuels have been a concern since Searchinger and Heimlich [87] first published a paper investigating LUC associated with bioenergy crop production and subsequent GHG emissions. Since that time, understanding of biofuel-associated LUC has improved greatly because better land use data are available and economic models for modeling LUC have been improved significantly. Recent LUC GHG results indicate that Searchinger and Heimlich [87] originally over-estimated LUC GHG emissions significantly [91].

Later, Searchinger and Heimlich [38] maintained that additional biomass for bioenergy is the key for bioenergy-derived GHG reductions. They further asserted that double counting of carbon occurs in analyses of bioenergy because the biomass additionality issue is not addressed. The concept of additionality for GHG reductions really asks what the counterfactual scenario is when bioenergy does not exist or exists only at current levels. The authors concluded that any dedicated biomass growth would result in double counting of biomass by assuming that biomass for bioenergy is grown on the lands where biomass would be grown anyway and with the same yields with no differences in land productivity of biomass between the two scenarios. Land biomass productivity differences could be caused by biomass yields, changes from single cropping to double cropping (or even triple cropping) because of bioenergy production [92], and the ability to use marginal/degraded lands for bioenergy production. Careful characterization of counterfactual scenarios and bioenergy scenarios has been conducted to address the biomass additionality issue. In fact, some analyses point to the need to compare biofuels to marginally produced, high-GHG-intensity fossil fuels [93].

According to Searchinger and Heimlich [38] phasing out bioenergy that uses crops or that otherwise makes dedicated use of land is a sound step toward a sustainable food future. However, U.S. farm acres for different crops illustrate that while corn acreage has increased in parallel with the build-up of the corn ethanol industry between 2004 and 2013, total principal crop acreage has remained fairly constant in the U.S. [94]. These observed trends are consistent with Taheripour and Tyner [95], who did observe crop shifting (e.g., wheat fields converted to corn agriculture) in the United States in this time period as a key mechanism for additional corn production. Another mechanism is likely the conversion of grasslands, wetlands, and other lands. The key point is that agricultural acreage in the United States has not significantly increased despite a dramatic biofuel boom. Additionally, the U.S. Renewable Fuel Standard states that biofuel feedstocks must come from land that was not forested before 2007. This provision limits the expansion of agricultural land into forested lands for biofuel production. In the case of woody feedstocks, the forests from which they derive must have been managed plantations before 2007.

Reduced deforestation and increased agricultural production can occur simultaneously in tropical forest frontiers, provided that land is available and policies promote the efficient use of already-cleared lands (intensification) while restricting deforestation. It remains uncertain whether government- and industry-led policies can contain deforestation if future market conditions favor another boom in agricultural expansion [96].

Academic studies using economic models have also found that biofuels can lead to reductions in lifecycle GHG emissions relative to conventional fuels [73,97]. Second and third generation biofuels have significant potential to reduce GHG emissions relative to conventional fuels because feedstocks can be produced using marginal land. Moreover, in the case of waste biomass, no additional agricultural production is required, and indirect market-mediated GHG emissions can be minimal if the wastes have no other productive uses.

Biofuel production and consumption, in and of itself, will not reduce GHG or conventional pollutant emissions, lessen petroleum imports, or alleviate pressure on exhaustible resources. Biofuel production and use must coincide with reductions in the production and use of fossil fuels for these benefits to accrue. These benefits would be mitigated if biofuel emissions and resource demands augment, rather than displace, those of fossil fuels.

Biofuel production and processing practices can also release GHGs. Fertilizer application releases nitrous oxide, a potent greenhouse gas. Most biorefineries operate using fossil fuels. Some research suggests that GHG emissions resulting from biofuel production and use, including those from indirect land use change, may be higher than those generated by fossil fuels, depending on the time horizon of the analysis [89,98].

Regarding non-GHG environmental impacts, research suggests that production of biofuel feedstocks, particularly food crops like corn and soy, could increase water pollution from nutrients, pesticides, and sediment [99]. Increases in irrigation and ethanol refining could deplete aquifers [99]. Air quality could also decline in some regions if the impact of biofuels on tailpipe emissions plus the additional emissions generated at biorefineries increases net conventional air pollution [99].

Biofuel co-products help mitigate the environmental consequences of expansion by the biofuel industry. For example, DDGS substitutes for both maize and soybean meal in livestock rations. This reduces the land use consequences of biofuel production and eases the demand for chemical inputs, such as fertilizers and pesticides, in crop production. By reporting only the gross usage of maize for ethanol, the implication was that all the maize going into ethanol production resulted in fuel ethanol [1]. According to the conventional assumption ethanol producers return a full one-third of the maize processed back to the feeding sector which is the difference between the gross and net volume of maize used for ethanol. However, in aggregate, a metric tonne of DDGS can replace, on average, 1.22 metric tonne of feed consisting of corn and soybean meal in the United States. In fact, the amount of feed (maize and soybean meal) replaced by the DDGS represents 38% a weight basis of the maize used in the associated ethanol production process for a given crop year. If co-products are taken into account the net use of feedstocks decline [18]. More complicated, but no less important, is the impact of DDGS on land use change and the GHG emissions associated with maize ethanol production. Most existing biofuel regulations significantly undervalue the contribution of DDGS when assessing the net GHG impacts of maize ethanol assuming that one metric tonne of DDGS replaces only one metric tonne of corn, with no substitution of soybean meal. The importance of DDGS is being undervalued by the regulatory agencies requiring a GHG assessment of ethanol. In the future accurate DDGS accounting is of increasing importance [1].

The proportion of global cropland used for biofuels is currently some 2% (30–35 million gross hectares) with wide differences among countries and regions. The review by [100] studies biofuel expansion between 2000 and 2010 in Brazil, the USA, Indonesia, Malaysia, China, Mozambique, South Africa plus 27 EU member states. In 2010, these countries produced 86 billion litres of ethanol and 15 billion litres of biodiesel representing 95% of global biofuel production. Land devoted to biofuel production was calculated at 32 million ha in 2010, an increase of 25 million ha as compared to 2000. According to [101] land required to produce biofuels grows from 30 million hectares in 2010 to 100 million hectares in 2050. Ignoring co-product generation in early biofuel impact assessments has led to an overestimation of land requirements, in most cases by 40% or more. The contribution of feed co-products is relatively high in the USA, China, and the EU due to the large share of cereals with high feed yields. It is low in Brazil where ethanol production is dominated by sugarcane which generates no feed co-products. Implications for land use will, however, also depend on the role of yield improvement [100]. Another report published in 2008 by CE Delft estimates that the use of co-products generated from rapeseed, soy, wheat and maize can reduce net land use by 11% to 25%. Biofuels produced from some feedstocks such as sugarcane, where nearly the entire product is used for producing biofuel, do not generate such co-products [102]. Co-products are supposed to be credited with the area of cropland required to produce the amount of feed they substitute. By adding co-products substituted for grains and oilseeds the land required for cultivation of feedstocks declines to 1.5% net land requirement of the global crop area. This substitution reduces the indirect land use and therefore the impact of indirect land use change and intensification is substantially [1].

In 2010 about 30–35 million gross hectares of feedstock was needed for fuel ethanol and biodiesel production. The proportion of global cropland used for biofuels is currently some 2% with wide differences among countries and regions. In the cases of grains and oilseeds, DDGS, CGF/CGM and oil cakes (mainly rapeseed and soybean cake/meal) substitute grain and soybean as feed. It means that not all the feedstocks used for bioethanol production should be subtracted from the supplies.

Based on the land-use efficiencies land use for biofuel production would need to increase from about 30–35 million hectares to around 100 million hectares in 2050. This corresponds to an increase from 2.5% of total arable land today to around 6% in 2050. This expansion would include some cropland, as well as pastures and currently unused land, the latter in particular for production of lignocellulosic biomass [2].

Algae as Animal Feed

Many studies have shown the suitability of algae as a potential animal feed and as a replacement for conventional protein sources such as soybean and fish meal. Unfortunately, the trend is to avoid using live algae due to their high cost and production difficulties. The use of micro-algae as animal feed is more recent. Algae biomass is a valuable feed supplement or substitute for conventional protein sources. The incorporation of algae into poultry rations offers the most promising prospect for their commercial use in animal feeding. Another growing market is the utilization of micro-algae in aquaculture [103]. In poultry, algae can be used as a partial replacement for conventional proteins with the incorporation of 51% [104]. Also, according to [105], they may serve as almost the sole protein source in laying hens, and in several countries, they are officially approved as chicken feed. In pigs ration, [106] assumed the incorporation of even 33%, without negative symptoms. According to another study the feed intake of the pigs on the algal meal diets was not affected until the third period of the experiment and suggests that the acceptance from a palatability point of view when added at 10% is not a major issue although the manipulation of the gut microbiome to algae meal is possibly of some concern although a larger number of piglets raised in commercial environments would be required to better define this as an issue [107]. However, it should be expected that the most suitable for feeding with algae are ruminants, because they are able to digest even an unprocessed algal cell wall [105]. Manufacturers of these organisms worldwide have recognized different potentials and therefore focused just on the food and feed industry [108]. Although, microalgae are eaten as a food in China and Chad and had been considered as a solution to the world's food shortage, their use on a global scale appears limited to health food and food supplements [109].

New Market Opportunities for Biofuel Co-Products

The biofuels industry has evolved rapidly over the last three decades with developments in processing techniques and an expansion of the range of plants and other natural energy sources being considered as feedstocks. On-farm application of the co-products, on which the viability of the industry depends, is often ahead of unbiased research to support its use. Much of the potential research identified as needed is concerned with co-product feeding value, the need for standardization of products from within an individual plant and between plants, and the search for new feedstocks together with safety standards. However, the economic importance of biofuel industrial production chains will increase in the coming years and will be a promising source of co-products that are useful for sustainable farming systems, biomaterials applications and crop disease management.

A potential market for DDGS and oilseed meals that biofuel producers may consider is the fertilizer/agriculture market. DDGS and oilseed meal co-products can be effective fertilizer sources for plants. In addition to containing most macronutrients and micronutrients needed to support plant growth, these co-products have low C:N ratios, which means that they are rapidly decomposable and can release nutrients to plants in a timely manner. This property is of particular interest to organic agriculture markets, where nutrient sources and fertilizers containing enough readily mineralizeable N to impact yield are scarce. Competition with animal feed markets has prevented widespread adoption of biofuel co-products as fertilizers. DDGS and many oilseed meals hold more value as an animal feed than as a fertilizer source, therefore animal producers are willing to pay for feed materials based on feed value instead of the lower fertilizer value. However, fertilizer markets may become more appealing to biofuel producers due to over-production of DDGS, high costs of oilseed meals, DDGS feed quality issues, and interest from organic growers in alternative nutrient-rich fertilizer sources [110,111,112].

The growth of biofuel industries caused the generation of a huge amount of under-valued co-products of different types including DDGS, bagasse, lignin, protein-rich meals and crude glycerol. In order to claim biofuel production as the sustainable technology of future, it is necessary to create value-addition to these co-products. This also provides solution to the emerging issues on the environmental impact of the accumulation of these co-products. Furthermore, the technological development related to the value-added uses of biofuel co-products in biomaterials applications is expected to create new ecofriendly products and strengthen the bioeconomy. Among these co-products, DDGS, CGM and soybean meal are traditionally used for animal feed applications. However, recent researches show the huge potential of these co-products for biomaterial applications (polymeric biocomposites, thermoplastic, biobased films for food packaging applications ect.). Crude glycerol co-produced in biodiesel industry has huge opportunities in biomaterial applications including chemicals/monomers, plasticizer, hydrogen generation, carbon source for bacterial growth and polyesters production. Lignin, the major co-product of lignocellulosic ethanol can be used as polymer blends/composites, adhesives, and carbon fibres. In addition to that, these biofuel co-products also create opportunity for the fabrication of various nanostructured materials for the high value applications [113].

Moreover, the use of co-products from the biofuel production chains, especially for crop disease management, is an under-explored area in the research community. In recent years, a relevant number of studies for the control of plant diseases have been carried out on bio-chars, oil-less seed meals, suppressive composts that are derived from agricultural waste and exhausted biomass, and steam-exploded liquid waste. The co-products of particular interest in crop protection are oil-less seed meals and glycerin derived from the biodiesel chain, steam-exploded liquid waste derived from a 2nd generation bioethanol chain, and charcoal obtained from the pyrolysis of plant biomass. Biofuel chain co-products have great potential but sometimes give inconsistent disease control, which limits their use in crop protection. Nevertheless, the benefits of biofuel chain co-products outweigh their drawbacks, but the impact of this approach on pathogen populations and disease suppression is often unpredictable. Future research should mainly focus on the development of biofuel chain co-products that enhance the activity of beneficial microbes and improvement of the effectiveness of biofuel chain co-products via applications of multicomponent bio-fungicides [114].

3. MATERIALS AND METHODS

The literature on the impacts of biofuel expansions is already substantial, however, the feed value of increasing biofuels co-products, which are supposed to be credited with the area of cropland required to produce the amount of feed they substitute have received much less attention.

The search platform Web of Knowledge and search engine Google Scholar are primarily used to collect the relevant literature. In addition, backward searches through bibliographies of academic studies and reviews as well as hand searching websites of academic projects and conferences on biofuel are also applied. Only literature in English is included in this paper so as to ensure accessibility. Since the rapid progress of this research filed, literature is also limited to the papers mainly published in or after 2010. The literature reviewed is selective and critical. Highly rated journals in scientific indexes are the preferred choice. We carefully select 114 papers which are considered as important or innovative studies, or comprehensive reviews offering us a big picture of biofuels and their co-products as livestock feed. The literature review is categorized into three topics:

1. Co-products of biofuels as livestock feed: biofuel's co-products refer to co-products that are generated during the process and used as animal feed
2. Economic, environmental and land use implications of biofuel's co-products. When co-products are used credits can be attributed to the biofuel production chain including economic viability, GHG emission savings and avoided land use. Impact on land use and GHG emissions are the main focuses the existing studies paid attention to.
3. Advanced biofuels: future production is influenced by the competition between conventional and advanced biofuels based on food crops and lignocellulosic feedstocks.

Amount of literature examined in this paper classified by topic:

Topic	Amount
Co-products	14
Land use	26
Economics	18
Environment	13
Bioenergy potential	12
Conventional biofuels	11
Advanced biofuels	20

4. CONCLUSIONS

Estimates on impacts of biofuel production often use models with limited ability to incorporate economic and environmental implications by ignoring generation of co-products from biofuel production. Co-product generation in early biofuel impact assessments was ignored leading to an overestimation of land requirements and GHG emissions. The output of feed co-products is relatively high in the USA, the EU and China due to the large share of grains used in ethanol production with high feed yields. It is low in Brazil where ethanol production is dominated by sugarcane which generates no feed co-products. Co-product yields are low for rapeseed and soybean used in the biodiesel industry. By economically displacing traditional feed ingredients co-products from biofuel production are an important and valuable component of the biofuels sector and the global feed market. Moreover, the return of co-products to the feed market has economic, land use and GHG emissions implications as well. Models used to evaluate biofuel policies should be enriched by incorporating more and better information on changes in land use, and economic and environmental implications of co-products. This information should be considered in discussions on food, feed versus fuel debate and land-use change caused by biofuel policies.

Recent years have seen a tremendous increase in the production of biofuels from agricultural commodities. Food-crop based feedstocks are expected to continue to dominate ethanol and biodiesel production over the coming decade. Growth in biofuel production has been accompanied by increased output of animal feed co-products from common biofuel processes. Globally, these feed co-products are growing in volume and importance. While the increased use of agricultural commodities for biofuels is generally expected to contribute to slightly higher input costs for certain livestock and poultry feeds, the impacts are expected to be modest and can be mitigated in part by increased substitution of co-products for traditional feedstuffs. Increased agricultural productivity has allowed the global supply of crops available for non-biofuel uses to continue to grow over the long term. The future use of agricultural crops for biofuel resulting in a small increase in livestock feed costs, which will be offset to some extent by the use of co-products as feed and by increases in crop yields over time. Feed co-product output is expected to grow more slowly in the coming years. However, a number of new and emerging technologies may change the composition and further improve the nutritional quality and utility of feed co-products. New technologies and practices promise to change the complexion of the ethanol co-products market in the years ahead.

Co-products are supposed to be credited with the area of cropland required to produce the amount of feed they substitute. If co-products are taken into account, the net use of feedstocks decline. By adding co-products substituted for grains and oilseeds the land required for cultivation of feedstocks declines from about 2% to 1.5% net land requirement of the global crop area. Moreover, it is important to include the co-products in GHG assessment, because of their potential impact on the overall emissions. Most existing biofuel regulations significantly undervalue the contribution of co-products when assessing the net land use and GHG impacts of biofuel production. In the future accurate co-products accounting is of increasing importance.

Growth in the use of agricultural commodities for biofuels is expected to continue through to 2020, but growth rates will slow in key producing countries as government-imposed limits on grain use for biofuels are reached and advanced biofuels capacity is expected to expand only slowly. The second reason for moderation in the growth in the use of agricultural commodities for biofuels is the expectation that future growth in biofuels production will primarily come from new feedstocks that currently have no or limited application in the animal feed market, such as perennial grasses (Miscanthus sp. and A.donax), agricultural residues, algae and other materials.

ACKNOWLEDGMENTS

This study was made with no grants received in support of our research work.

AUTHOR CONTRIBUTIONS

All authors collected and processed data. Sections of paper were written by different authors: biofuels production by József Popp, substitution of traditional animal feed with co-products by Mónika Harangi-Rákos, impact of biofuel production on land use by Péter Balogh, giant reed as perspective feedstock for biofuels or biogas production by Gabriella Antal and biogas production based on algae by Zoltán Gabnai and Attila Bai. All authors participated in the study design. All authors read and approved the final version of the manuscript.

CONFLICTS OF INTEREST

The authors declare no conflict of interest.

REFERENCES

1. Popp, J.; Lakner, Z.; Harangi-Rakos, M.; Fari, M. The effect of bioenergy expansion: Food, energy, and environment. Renew. Sustain. Energy Rev. **2014**, 32, 559–578.
2. International Energy Agency. The Renewable Energy 2013. Medium-Term Market Report. A Growing Role for Renewables in the Energy Mix; International Energy Agency: Paris, France, 2013; p. 13.
3. World Bioenergy Association. Global Bioenergy Statistics 2014; WBA: Stockholm, Sweden, 2014.
4. United States Department of Agriculture. Agricultural Projections to 2024; Interagency Agricultural Projections Committee: Washington, DC, USA, 2015; p. 91.
5. OECD/FAO. Agricultural Outlook; 2015; Avaliable online: http://dx.doi.org/10.1787/data-00736-en (assessed on 10 December 2015).
6. International Energy Agency. World Energy Outlook 2015; International Energy Agency: Paris, France, 2015; p. 200.
7. Licht, F.O. World Ethanol and Biofuels Report; Agra Informa: London, UK, 2013.
8. Elobeid, A.E.; Carriquiry, M.A.; Fabiosa, J.F. Land-use change and greenhouse gas emissions in the fapri-card model system: Addressing bias and uncertainty. Clim. Chang. Econ. **2012**, 3.
9. Khanna, M.; Zilberman, D. Modeling the land-use and greenhouse-gas implications of biofuels. Clim. Chang. Econ. **2012**, 3.
10. Witzke, P.; Banse, M.; Gömann, H.; Heckelei, T.; Breuer, T.; Mann, S.; Kempen, M.; Adenäuer, M.; Zintl, A. Modelling of Energy-Crops in Agricultural Sector Models—A Review of Existing Methodologies; Joint Research Centre—Institute for Prospective Technological Studies Luxembourg: Ispra, Italy, 2008; p. 59.

11. Golub, A.A.; Hertel, T.W. Modeling land-use change impacts of biofuels in the gtap-bio framework. Clim. Chang. Econ. **2012**, 3.

12. Beach, R.H.; Zhang, Y.W.; McCarl, B.A. Modeling bioenergy, land use, and ghg emissions with fasomghg: Model overview and analysis of storage cost implications. Clim. Chang. Econ. **2012**, 3.

13. Makkar, H.P.S. Biofuel Co-Products as Livestock Feed—Opportunities and Challenges; FAO: Rome, Italy, 2012; p. 533.

14. International Energy Agency. World Energy Outlook 2012; International Energy Agency: Paris, France, 2012; p. 690.

15. Smeets, E.M.W.; Faaij, A.P.C.; Lewandowski, I.M.; Turkenburg, W.C. A bottom-up assessment and review of global bio-energy potentials to 2050. Prog. Energy Combust. **2007**, 33, 56–106.

16. International Energy Agency. A Sustainable and Reliable Energy Source Main Report; International Energy Agency: Paris, France, 2009; p. 108.

17. Dornburg, V.; van Vuuren, D.; van de Ven, G.; Langeveld, H.; Meeusen, M.; Banse, M.; van Oorschot, M.; Ros, J.; van den Born, G.J.; Aiking, H.; et al. Bioenergy revisited: Key factors in global potentials of bioenergy. Energy Environ. Sci. **2010**, 3, 258–267.

18. Intergovernmental Panel on Climate Change. Special Report on Renewable Energy Sources and Climate Change Mitigation; Cambridge University Press: Cambridge, UK, 2011; p. 246.

19. Haberl, H.; Beringer, T.; Bhattacharya, S.C.; Erb, K.H.; Hoogwijk, M. The global technical potential of bio-energy in 2050 considering sustainability constraints. Curr. Opin. Environ. Sustain. **2010**, 2, 394–403.

20. Global Energy Assessment. Global Energy Assessment—Toward a Sustainable Future; Cambridge University Press: Cambridge, UK, 2012; p. 113.

21. Haberl, H.; Erb, K.H.; Krausmann, F.; Gaube, V.; Bondeau, A.; Plutzar, C.; Gingrich, S.; Lucht, W.; Fischer-Kowalski, M. Quantifying and mapping the human appropriation of net primary production in earth's terrestrial ecosystems. PNAS **2007**, 104, 12942–12945.

22. Larcher, W. Physiological Plant Ecology: Ecophysiology and Stress Physiology of Functional Groups; Springer Science & Business Media: Berlin, Germany, 2003; p. 513.

23. Saugier, B.; Roy, J.; Mooney, H.A. Estimations of global terrestrial productivity: Converging toward a single number. In Terrestrial Global Productivity; Academic Press: San Diego, CA, USA, 2001; pp. 543–557.

24. Food and Agriculture Organization. Faostat Database; Economic and Social Development Department: Rome, Italy, 2011.

25. International Energy Agency. Energy Technology Perspectives 2010. Scenarios & Strategies to 2050; International Energy Agency: Paris, France, 2010; p. 706.

26. Paulová, L.; Melzoch, K.; Rychtera, M.; Patáková, P. Production of 2nd generation of liquid biofuels. In Liquid, Gaseous and Solid Biofuels: Conversion Techniques; Fang, Z., Ed.; InTech Open Access Publisher: Rijeka, Croatia, 2013; pp. 47–78.

27. Alshaal, T.; Domokos-Szabolcsy, E.; Marton, L.; Czako, M.; Katai, J.; Balogh, P.; Elhawat, N.; El-Ramady, H.; Fari, M. Phytoremediation of bauxite-derived red mud by giant reed. Environ. Chem. Lett. **2013**, 11, 295–302.

28. El-Ramady, H.R.; Abdalla, N.; Alshaal, T.; Fári, M.; Prokisch, J.; Pilon-Smits, E.A.H.; Domokos-Szabolcsy, É. Selenium phytoremediation by giant reed. In Hydrogen Production and Remediation of Carbon and Pollutants; Lichtfouse, E., Schwarzbauer, J., Robert, D., Eds.; Springer International Publishing: Cham, Switzerland, 2015; pp. 133–198.

29. Pandey, V.C.; Bajpai, O.; Singh, N. Energy crops in sustainable phytoremediation. Renew. Sustain. Energy Rev. **2016**, 54, 58–73.

30. Jones, M.B.; Finnan, J.; Hodkinson, T.R. Morphological and physiological traits for higher biomass production in perennial rhizomatous grasses grown on marginal land. Glob. Chang. Biol. Bioenergy **2015**, 7, 375–385.

31. Ge, X.M.; Xu, F.Q.; Vasco-Correa, J.; Li, Y.B. Giant reed: A competitive energy crop in comparison with miscanthus. Renew. Sustain. Energy Rev. **2016**, 54, 350–362.

32. Giessow, J.; Casanova, J.; Leclerc, R.; MacArthur, R.; Fleming, G.; Giessow, J. Arundo donax (Giant Reed): Distribution and Impact Report; State Water Resources Control Board California Invasive Plant Council (Cal-IPC): California, CA, USA, 2011; pp. 1–240.

33. Maceda-Veiga, A.; Basas, H.; Lanzaco, G.; Sala, M.; Sostoa, A.; Serra, A. Impacts of the invader giant reed (Arundo donax) on riparian

habitats and ground arthropod communities. Biol. Invasions **2016**, 18, 731–749.

34. Nackley, L.L.; Kim, S.-H. A salt on the bioenergy and biological invasions debate: Salinity tolerance of the invasive biomass feedstock Arundo donax. GCB Bioenergy **2015**, 7, 752–762.

35. International Energy Agency. Renewable Energy. Medium-Term Market Report 2015. Market Analysis and Forecasts to 2020; International Energy Agency: Paris, France, 2015; p. 14.

36. Kretschmer, B.; Allen, B.; Kieve, D.; Smith, C. The Sustainability of Advanced Biofuels in the EU: Assessing the Sustainability of Wastes, Residues and Other Feedstocks Set out in the European Commission's Proposal on indirect Land Use Change (ILUC); Institute for European Environmental Policy (IEEP): London, UK, 2013; p. 37.

37. Renewable Energy Policy Network for the 21st Century. Renewables 2015, Global Status Report; Paris, France, 2015; p. 251.

38. Searchinger, T.; Heimlich, R. Avoiding Bioenergy Competition for Food Crops and Land; World Resources Institute: Washington, DC, USA, 2015; p. 44.

39. Wang, M.; Dunn, J.B. Comments on Avoiding Bioenergy Competition for Food Crops and Land by Searchinger and Heimlich; Argonne National Laboratory: Argonne, IL, USA, 2015; pp. 1–9.

40. Lynd, L.R.; Larson, E.; Greene, N.; Laser, M.; Sheehan, J.; Dale, B.E.; McLaughlin, S.; Wang, M. The role of biomass in america's energy future: Framing the analysis. Biofuels Bioprod. Biorefin. **2009**, 3, 113–123.

41. Cai, X.M.; Zhang, X.A.; Wang, D.B. Land availability for biofuel production. Environ. Sci. Technol. **2011**, 45, 334–339.

42. Perlack, R.D.; Eaton, L.M.; Turhollow, A.F., Jr.; Langholtz, M.H.; Brandt, C.C.; Downing, M.E.; Graham, R.L.; Wright, L.L.; Kavkewitz, J.M.; Shamey, A.M. Us Billion-Ton Update: Biomass Supply for a Bioenergy and Bioproducts Industry; U.S. Department of Energy: Oak Ridge, Tennessee, TN, USA, 2011; p. 227.

43. Werling, B.P.; Dickson, T.L.; Isaacs, R.; Gaines, H.; Gratton, C.; Gross, K.L.; Liere, H.; Malmstrom, C.M.; Meehan, T.D.; Ruan, L.L.; et al. Perennial grasslands enhance biodiversity and multiple ecosystem services in bioenergy landscapes. PNAS **2014**, 111, 1652–1657.

44. Stolarski, M.J.; Krzyzaniak, M.; Luczyski, M.; Zaluski, D.; Szczukowski, S.; Tworkowski, J.; Golaszewski, J. Lignocellulosic biomass from short rotation woody crops as a feedstock for second-generation bioethanol production. Ind. Crop Prod. **2015**, 75, 66–75.

45. Allwright, M.R.; Taylor, G. Molecular breeding for improved second generation bioenergy crops. Trends Plant Sci. **2016**, 21, 43–54.

46. Triana, F.; Nassi o Di Nasso, N.; Ragaglini, G.; Roncucci, N.; Bonari, E. Evapotranspiration, crop coefficient and water use efficiency of giant reed (Arundo donax L.) and miscanthus (Miscanthus × giganteus greef et deu.) in a mediterranean environment. GCB Bioenergy **2015**, 7, 811–819.

47. Testa, R.; Foderà, M.; Di Trapani, A.M.; Tudisca, S.; Sgroi, F. Giant reed as energy crop for southern italy: An economic feasibility study. Renew. Sustain. Energy Rev. **2016**, 58, 558–564.

48. Alshaal, T.; Domokos-Szabolcsy, E.; Marton, L.; Czako, M.; Katai, J.; Balogh, P.; Elhawat, N.; El-Ramady, H.; Gerocs, A.; Fari, M. Restoring soil ecosystems and biomass production of Arundo donax L. Under microbial communities-depleted soil. Bioenergy Res. **2014**, 7, 268–278.

49. Alshaal, T.; Elhawat, N.; Domokos-Szabolcsy, É.; Kátai, J.; Márton, L.; Czakó, M.; El-Ramady, H.; Fári, M.G. Giant reed (Arundo donax L.): A green technology for clean environment. In Phytoremediation; Ansari, A.A., Gill, S.S., Gill, R., Lanza, G.R., Newman, L., Eds.; Springer International Publishing: Cham, Switzerland, 2014; Volume 1, pp. 3–20.

50. Elhawat, N.; Alshaal, T.; Domokos-Szabolcsy, E.; El-Ramady, H.; Marton, L.; Czako, M.; Katai, J.; Balogh, P.; Sztrik, A.; Molnar, M.; et al. Phytoaccumulation potentials of two biotechnologically propagated ecotypes of arundo donax in copper-contaminated synthetic wastewater. Environ. Sci. Pollut. Res. **2014**, 21, 7773–7780.

51. Elhawat, N.; Alshaal, T.; Domokos-Szabolcsy, E.; El-Ramady, H.; Antal, G.; Marton, L.; Czako, M.; Balogh, P.; Fari, M. Copper uptake efficiency and its distribution within bioenergy grass giant reed. Bull. Environ. Contam. Toxicol. **2015**, 95, 452–458.

52. Rabemanolontsoa, H.; Saka, S. Comparative study on chemical composition of various biomass species. RSC Adv. **2013**, 3, 3946–3956.

53. Franscisco, L.; Carlos, G.J.; Antonie, P.; Javier, F.M.; Minerva, A.M.Z.; Gil, G. Chemical and energetic characterization of species with a high-biomass production: Fractionation of their components. Environ. Prog. Sustain. **2010**, 29, 499–509.

54. Shatalov, A.A.; Pereira, H. Xylose production from giant reed (Arundo donax L.): Modeling and optimization of dilute acid hydrolysis. Carbohyd. Polym. **2012**, 87, 210–217.

55. Komolwanich, T.; Tatijarern, P.; Prasertwasu, S.; Khumsupan, D.; Chaisuwan, T.; Luengnaruemitchai, A.; Wongkasemjit, S. Comparative potentiality of kans grass (Saccharum spontaneum) and giant reed (Arundo donax) as lignocellulosic feedstocks for the release of monomeric sugars by microwave/chemical pretreatment. Cellulose **2014**, 21, 1327–1340.

56. Corno, L.; Pilu, R.; Adani, F. Arundo donax L.: A non-food crop for bioenergy and bio-compound production. Biotechnol. Adv. **2014**, 32, 1535–1549.

57. Silva, C.F.L.E.; Schirmer, M.A.; Maeda, R.N.; Barcelos, C.A.; Pereira, N. Potential of giant reed (Arundo donax L.) for second generation ethanol production. Electron. J. Biotechnol. **2015**, 18, 10–15.

58. Darzins, A.; Pienkos, P.; Edye, L. Current Status and Potential for Algal Biofuels Production; National Renewable Energy Laboratory: Golden, CO, USA, 2010; pp. 1–131.

59. International Energy Agency. Algae—The Future for Bioenergy? Summary and Conclusions from the Iea Bioenergy Exco64 Workshop; International Energy Agency: Paris, France, 2010; p. 16.

60. Sikes, K.; McGill, R.; van Walwijk, M. Algae as a Feedstock for Biofuels: An Assessment of the State of Technology and Opportunities.Final Report; IEA-Advanced Motor Fuels Agreement, Annex 34, Subtask 2 USA; 2011; pp. 1–23.

61. Diltz, R.; Pullammanappallil, P. Biofuels from algae. In Liquid, Gaseous and Solid Biofuels: Conversion Techniques; Fang, Z., Ed.; InTech Open Access Publisher: Rijeka, Croatia, 2013; pp. 431–449.

62. Ndimba, B.K.; Ndimba, R.J.; Johnson, T.S.; Waditee-Sirisattha, R.; Baba, M.; Sirisattha, S.; Shiraiwa, Y.; Agrawal, G.K.; Rakwal, R. Biofuels as a sustainable energy source: An update of the applications of proteomics in bioenergy crops and algae. J. Proteom. **2013**, 93, 234–244.

63. Huntley, M.E.; Johnson, Z.I.; Brown, S.L.; Sills, D.L.; Gerber, L.; Archibald, I.; Machesky, S.C.; Granados, J.; Beal, C.; Greene, C.H. Demonstrated large-scale production of marine microalgae for fuels and feed. Algal Res. **2015**, 10, 249–265.

64. Zhou, W.G.; Chen, P.; Min, M.; Ma, X.C.; Wang, J.H.; Griffith, R.; Hussain, F.; Peng, P.; Xie, Q.L.; Li, Y.; et al. Environment-enhancing algal biofuel production using wastewaters. Renew. Sustain. Energy Rev. **2014**, 36, 256–269.

65. Kligerman, D.C.; Bouwer, E.J. Prospects for biodiesel production from algae-based wastewater treatment in brazil: A review. Renew. Sustain. Energy Rev. **2015**, 52, 1834–1846.

66. Hoh, D.; Watson, S.; Kan, E. Algal biofilm reactors for integrated wastewater treatment and biofuel production: A review. Chem. Eng. J. **2016**, 287, 466–473.

67. Gnansounou, E.; Raman, J.K. Life cycle assessment of algae biodiesel and its co-products. Appl. Energy **2016**, 161, 300–308.

68. High Level Panel of Experts on Food Security and Nutrition. Biofuels and Food Security; HLPE: Rome, Italy, 2013; pp. 1–131.

69. Association, R.F. Fueling a Nation Feeding the World. The role of the U.S. Ethanol Industry in Food and Feed Production; Washington, DC, USA, 2014; pp. 1–11.

70. Wisner, R. November monthly renewable energy report. Distillers grains balance sheet. Agricultural Marketing Resource Center: Ames, Iowa, IA, USA, 2015.

71. European Commission. Directive 2009/28/ec of the european parliament and of the council of 23 April 2009 on the promotion of the use of energy from renewable sources and amending and subsequently repealing directives 2001/77/ec and 2003/30/ec, (oj l 140/16). In Directive 2009/28/EC; European Commission: Brussels, Belgium, 2009; pp. 16–62.

72. Organization for Economic Cooperation and Development. Biofuel Support Policies: An Economic Assessment; OECD Publishing: Paris, France, 2008.

73. Huang, H.X.; Khanna, M.; Onal, H.; Chen, X.G. Stacking low carbon policies on the renewable fuels standard: Economic and greenhouse gas implications. Energy Policy **2013**, 56, 5–15.

74. Lynd, L.R.; Sow, M.; Chimphango, A.F.A.; Cortez, L.A.B.; Cruz, C.H.B.; Elmissiry, M.; Laser, M.; Mayaki, I.A.; Moraes, M.A.F.D.; Nogueira, L.A.H.; et al. Bioenergy and african transformation. Biotechnol. Biofuels **2015**, 8, 18.

75. Dale, B.E.; Anderson, J.E.; Brown, R.C.; Csonka, S.; Dale, V.H.; Herwick, G.; Jackson, R.D.; Jordan, N.; Kaffka, S.; Kline, K.L.; et al. Take a closer look: Biofuels can support environmental, economic and social goals. Environ. Sci. Technol. **2014**, 48, 7200–7203.

76. Trostle, R.; Marti, D.; Rosen, S.; Westcott, P. Why Have Food Commodity Prices Risen Again; Economic Research Service: Washington, DC, USA, 2011; pp. 1–29.

77. Zhang, W.; Yu, E.A.; Rozelle, S.; Yang, J.; Msangi, S. The impact of biofuel growth on agriculture: Why is the range of estimates so wide? Food Policy **2013**, 38, 227–239.

78. De Gorter, H.; Drabik, D. Policy update: The effect of biofuel policies on food grain commodity prices. Biofuels **2012**, 3, 21–24.

79. De Gorter, H.; Drabik, D. Policy update: Biofuel policies and grain crop price volatility. Biofuels **2012**, 3, 111–113.

80. De Gorter, H.; Drabik, D.; Just, D.R. Biofuel policies and food grain commodity prices 2006–2012: All boom and no bust? AgBioForum **2013**, 16, 1–13.

81. Durham, C.; Davies, G.; Bhattacharyya, T. Can Biofuels Policy Work for Food Security? An Analytical Paper for Discussion; UK Department for Environment, Food and Rural Affairs (DEFRA): Devon, UK, 2012; pp. 1–29.

82. Hamelinck, C. Biofuels and Food Security. Risks and Opportunities; ECOFYS: Utrecht, The Netherlands, 2013; pp. 1–23.

83. Taheripour, F.; Hertel, T.W.; Tyner, W.E.; Beckman, J.F.; Birur, D.K. Biofuels and their by-products: Global economic and environmental implications. Biomass Bioenergy **2010**, 34, 278–289.

84. Irwin, S.; Good, D. Ethanol production profits: The risk from lower prices of distillers grains. Farmdoc Dly. **2015**, 5, 46.

85. European Commission. Directive 2009/30/ec of the european parliament and of the council of 23 April 2009 amending directive 98/70/ec as regards the specification of petrol, diesel and gas-oil and introducing a mechanism to monitor and reduce greenhouse

gas emissions and amending council directive 1999/32/ec as regards the specification of fuel used by inland waterway vessels and repealing directive 93/12/eec. In Directive 2009/30/EC; European Commission: Brussels, Belgium, 2009; pp. 88–113.

86. OECD/FAO. Agricultural Outlook 2009–2018; Paris, France, 2009; pp. 1–95.

87. Searchinger, T.; Heimlich, R.; Houghton, R.A.; Dong, F.X.; Elobeid, A.; Fabiosa, J.; Tokgoz, S.; Hayes, D.; Yu, T.H. Use of us croplands for biofuels increases greenhouse gases through emissions from land-use change. Science **2008**, 319, 1238–1240.

88. Fargione, J.; Hill, J.; Tilman, D.; Polasky, S.; Hawthorne, P. Land clearing and the biofuel carbon debt. Science **2008**, 319, 1235–1238.

89. Melillo, J.M.; Reilly, J.M.; Kicklighter, D.W.; Gurgel, A.C.; Cronin, T.W.; Paltsev, S.; Felzer, B.S.; Wang, X.D.; Sokolov, A.P.; Schlosser, C.A. Indirect emissions from biofuels: How important? Science **2009**, 326, 1397–1399.

90. Fritsche, U.R.; Wiegmann, K. Indirect Land Use Change and Biofuels; Directorate General For Internal Policies, Policy Department A: Economic and Scientific Policy Environment, Public Health and Food Safety: Brussels, Belgium, 2011.

91. Dunn, J.B.; Mueller, S.; Kwon, H.Y.; Wang, M.Q. Land-use change and greenhouse gas emissions from corn and cellulosic ethanol. Biotechnol. Biofuels **2013**, 6.

92. Babcock, B.A.; Iqbal, Z. Using Recent Land Use Changes to Validate Land Use Change Models; Center for Agricultural and Rural Development: Ames, IA, USA, 2014; pp. 1–34.

93. Van den Bos, A.; Hamelinck, C. Greenhouse Gas Impact of Marginal Fossil Fuel Use; ECOFYS: Utrecht, The Netherlands, 2014; pp. 1–29.

94. Wang, M.J.; Dunn, B.; Mueller, S.; Qin, Z.; Tyner, W.; Goodwin, B. Comments on Ethanol's Broken Promise by the Environmental Working Group (May 2014); Argonne National Laboratory: Argonne, IL, USA, 2014; pp. 1–13.

95. Taheripour, F.; Tyner, W.E. Biofuels and land use change: Applying recent evidence to model estimates. Appl. Sci. **2013**, 3, 14–38.

96. Macedo, M.N.; DeFries, R.S.; Morton, D.C.; Stickler, C.M.; Galford, G.L.; Shimabukuro, Y.E. Decoupling of deforestation and soy

production in the southern amazon during the late 2000s. PNAS **2012**, 109, 1341–1346.

97. Hertel, T.W.; Golub, A.A.; Jones, A.D.; O'Hare, M.; Plevin, R.J.; Kammen, D.M. Effects of us maize ethanol on global land use and greenhouse gas emissions: Estimating market-mediated responses. Bioscience **2010**, 60, 223–231.

98. Mosnier, A.; Havlik, P.; Valin, H.; Baker, J.; Murray, B.; Feng, S.; Obersteiner, M.; McCarl, B.A.; Rose, S.K.; Schneider, U.A. Alternative us biofuel mandates and global ghg emissions: The role of land use change, crop management and yield growth. Energy Policy **2013**, 57, 602–614.

99. National Research Council. Renewable Fuel Standard: Potential Economic and Environmental Effects of U.S. Biofuel Policy; The National Academies Press: Washington, DC, USA, 2011; p. 250.

100. Langeveld, J.W.A.; Dixon, J.; van Keulen, H.; Quist-Wessel, P.M.F. Analyzing the effect of biofuel expansion on land use in major producing countries: Evidence of increased multiple cropping. Biofuel Bioprod. Biorefin. **2014**, 8, 49–58.

101. International Energy Agency. Technology Roadmap Biofuels for Transport; International Energy Agency: Paris, France, 2011; p. 52.

102. Croezen, H.; Brouwer, F. Estimating Indirect Land Use Impacts from by-Products Utilization; CE Delft: Delft, The Netherlands, 2008; pp. 1–31.

103. Becker, E.W. Micro-algae as a source of protein. Biotechnol. Adv. **2007**, 25, 207–210.

104. Spolaore, P.; Joannis-Cassan, C.; Duran, E.; Isambert, A. Commercial applications of microalgae. J. Biosci. Bioeng. **2006**, 101, 87–96.

105. Gouveia, L.; Batista, A.P.; Sousa, I.; Raymundo, A.; Bandarra, N.M. Microalgae in novel food products. In Food Chemistry Research Developments; Papadoupoulos, K., Ed.; Nova Science Publishers: New York, NY, USA, 2008; pp. 75–112.

106. Yap, T.N.; Wu, J.F.; Pond, W.G.; Krook, L. Feasibility of feeding spirulina-maxima, arthrospira-platensis or chlorella sp to pigs weaned to a dry diet at 4 to 8 days of age. Nutr. Rep. Int. **1982**, 25, 543–552.

107. Henman, D.; Australia, R. Evaluation of Algal Meal as an Energy and Protein Source in Pig Diets 4a-102. Report Prepared for the Co-

Operative Research Centre for High Integrity Australian Pork; Australian Government's Cooperative Research Centres: Corowa, New South Wales, Australia, 2012; pp. 1–12.

108. Kovač, D.J.; Simeunović, J.B.; Babić, O.B.; Mišan, A.Č.; Milovanović, I.L. Algae in food and feed. Food Feed Res. **2013**, 40, 21–32.

109. Milledge, J.J. Commercial application of microalgae other than as biofuels: A brief review. Rev. Environ. Sci. Biotechnol. **2011**, 10, 31–41.

110. Nelson, K.A.; Motavalli, P.P.; Smoot, R.L. Utility of dried distillers grain as a fertilizer source for corn. J. Agric. Sci. **2009**, 1, 3–12.

111. Abe, S.S.; Hashimoto, S.; Umezane, T.; Yamaguchi, T.; Yamamoto, S.; Yamada, S.; Endo, T.; Nakata, N. Agronomic and environmental performance of rapeseed oilcake in the lowland rice farming of japan. Commun. Soil Sci. Plant Anal. **2010**, 41, 1944–1958.

112. Snyder, A.J.; Johnson-Maynard, J.L.; Morra, M.J. Nitrogen mineralization in soil incubated with n-15-labeled brassicaceae seed meals. Appl. Soil Ecol. **2010**, 46, 73–80.

113. Vivekanandhan, S.; Mohanty, A.; Misra, M.; Zarrinbakhsh, N. Coproducts of biofuel industries in value-added biomaterials uses: A move towards a sustainable bioeconomy. In Liquid, Gaseous and Solid Biofuels: Conversion Techniques; Fang, Z., Ed.; InTech Open Access Publisher: Rijeka, Croatia, 2013; pp. 491–541.

114. De Corato, U.; Pane, C.; Bruno, G.L.; Cancellara, F.A.; Zaccardelli, M. Co-products from a biofuel production chain in crop disease management: A review. Crop Prot. **2015**, 68, 12–26.

3
CHAPTER

BALLERINI, D BIOFUELS: MEETING THE ENERGY AND ENVIRONMENTAL CHALLENGES OF THE TRANSPORTATION SECTOR

Arne Gröngröft and Franziska Müller-Langer*

Biofuels: Meeting the Energy and Environmental Challenges of the Transportation Sector

With the book *Biofuels, Meeting the Energy and Environmental Challenges of the Transportation Sector*, Daniel Ballerini published a comprehensive overview of biofuel production and markets. It is a revised and translated version of the book *Les Biocarburants - État des lieux, perspectives et enjeux du dévelopement*, which was first published in 2006 but has been only available so far in French. Changes in the world energy context, the increasing awareness of the environmental stakes and the development of research into the production of second- and third-generation biofuels have created a need for a new book to be written. The back cover of the book proclaims that it provides a detailed state-of-the-art description of first-generation biofuel production technologies. Furthermore, it describes the new second- generation pathways that use lignocellulosic biomass as a raw material and have been found to have industrial applications. The book also provides a technical update of the algae-to-energy pathway (third generation) and the production of methane and hydrogen via biochemical pathways.

To comply with this aim, Daniel Ballerini gathered 22 experts in the field to contribute their knowledge as authors. Many of the authors are affiliated with the French IFP Énergies nouvelles, which is surely a reason why the elaborations on biofuels are well connected to general issues in the petrol industry. Figures and tables adequately complement the text. Each chapter is followed directly by a list of references, of which only few are French publications. A five-page index may serve for navigation in the book.

The book's seven chapters follow a pragmatic structure. The first chapter is an introduction to the topic. Chapters two and three deal with current biofuels for spark ignition engines and diesel engines, respectively. Chapter four gives an introduction to 'lignocellulosic biomass resources'. Their conversion into biofuels via thermochemical and biochemical pathways is treated in chapters five and six, respectively. Finally, chapter seven covers 'other biomass-to-energy biochemical pathways'.

Chapter one provides a comprehensive introduction to the controversially discussed topic of biofuels. It starts with a short historical background of biofuels, beginning in the 1960s, which were envisaged to be the main energy carriers of choice in the early days of motorised transport. It then gives a good overview of the competitiveness and different interests and concerns of many of the stakeholders that are involved in biofuel production. Amongst the topics discussed are safety issues in handling and production, environmental and ecological impacts of biomass production and biofuel use. The visions of the refining industry as well as the car manufacturers towards biofuels are also elaborated. Moreover, a short overview of the situation in the mid-2000s is given for the most important regions, such as America and Europe, with some insight into other global regions.

Chapter two addresses the production of ethanol from sugar and amylaceous crops as well as ether fuels. A brief overview of current ethanol production practices is followed by energy balances and a list of improvement options for these processes. To find a whole section about ethers, such as ETBE, MTBE, TAME and ETAE, in a book about biofuels expands the horizon to the possibilities for an incorporation of bio-based compounds like ethanol into the petrol industry. Moreover, the most important industrial players and the state of global ethanol production are briefly addressed, however, only until 2008. Sections about fuel standards, impacts on engine behaviour and conversion of ethanol into other chemicals and fuels will finally lead us to the environmental and economic aspects of ethanol and ethers.

Chapter three discusses the production, utilisation and environmental and economic aspects of the so-called first-generation biofuels for diesel engines. Starting with an introduction to vegetable oils as a raw material, biodiesel production via transesterification is mainly covered. This is partnered with a short introduction to diesel fuel production via hydrotreatment and hydroisomerisation. Again, some of the most important industrial players and the state of global ethanol production are briefly addressed, unfortunately only until 2008. The direct combustion of plant oils in diesel engines is also presented but with the conclusion that manufacturers do not tolerate the resulting risks. A rarity in the main chapter is the subchapter on the use of biofuels for aviation, which addresses different fuel options.

Chapter four sets the stage for the following chapters by elaborating on the different types of lignocellulosic biomass, their composition, technical potentials and accessibility. Most advantages and disadvantages of second-generation biofuels can be deducted from their feedstock characteristics.

Chapters five and six are dedicated to the production of the so-called second-generation biofuels.

Chapter five deals with thermochemical conversion such as pyrolysis, torrefaction, gasification and hydrothermal conversion. In logical succession, it also contains the synthesis of a number of liquid biofuels and hydrogen from bio-based syngas. The gasification section includes a lot of issues linked to synthesis focussing on Fischer-Tropsch synthesis. The quite important section on gas treatment is less comprehensive and does not adequately include the many different options for gas cleaning and treatment to syngas. For hydrothermal processes, only HTU® is considered while other alternatives of hydrothermal carbonisation and liquefaction are also part of research and development (R&D) in Europe. The section on synfuel covers the most important syntheses to liquid fuels (methanol, DME, Fischer-Tropsch) and hydrogen but does not even mention the internationally relevant gaseous fuel, synthetic natural gas via syngas methanation (e.g. demo plant in Güssing). For the mentioned project references, the past few years have shown a dynamic development (e.g. some demo plants have not yet run, such as in Freiberg/Germany).

Chapter six deals with both the biochemical conversion to ethanol and the biochemical conversion using acetonobutylic fermentation (ABE). For the ethanol part, the many options of how to disintegrate, hydrolyse and ferment lignocellulose are shortly presented. Some of the biochemical issues of cellulase production and pentose fermentation are also highlighted. A closer look at the recent developments in this sector, with numerous demo plants in operation and (semi) commercial scale plants announced, would have been perfect. Finally, a historical and technical overview of ABE (acetone-butanol-ethanol) fermentation is given.

Chapter seven deals with the fermentative options for the production of biomethane or hydrogen. While biomethane via biogas upgrading is a commercial technology implemented across Europe, the hydrogen options are on an early R&D level. Moreover, algae production is included, which also addresses different algae options and the extraction of algae oils.

Finally, it can be concluded that this book stands out from a number of published books with similar titles in the last years by being more than a collection of stand-alone articles. By treating almost all of the relevant biofuel options, it provides a valuable background on technical, environmental and economic aspects for anybody who is interested in the complex topic of biofuels. Ballerini's *Biofuels* is well structured, informative and should prove to be a good reference book for both students and professionals from the industry and research, dealing with alternative fuels.

DECLARATIONS

Acknowledgements

The authors would like to thank very much Dagmar Fiedler from the Helmholtz Centre for Environmental Research – UFZ, Department of Environmental Microbiology for giving the opportunity to review the above mentioned book.

Competing interests

Both authors declare that they have no competing interests.

Authors' contributions

FML conceived the book review and participated in its design and coordination. AG carried out the major part of the review and drafted the manuscript. Both authors read and approved the final manuscript.

4

CHAPTER

AIR QUALITY AND BIOFUELS

S. Prasad[1] and M.S. Dhanya

[1] Division of Environmental Sciences,Indian Agricultural Research Institute, New Delhi, India

1. INTRODUCTION

The energy sector has played a crucial role in the context of the global economy as well as the socio-economic development. The world energy consumption is growing at the rate of 2.3% per year. The Energy Information Administration estimated that the primary sources of energy consisted of petroleum 36.0%, coal 27.4% and natural gas 23.0% amounting to 86.4% share for fossil fuels in primary energy consumption in the world (EIA, 2010). Fossil fuel consumption is the largest contributor to air pollution, greenhouse gas emissions and the environmental impacts with a large endowment of coal and has an energy system that is highly carbon intensive. The combustion of fossil fuel releases VOCs, nitrogen oxides (NOx), carbon monoxide (CO) and particulate matter (PM). The combination of VOCs and NOx with sunlight further results in the formation of tropospheric ozone, the main component of smog. The burning of fossil fuels produces around 21.3 billion tonnes (21.3 giga tonnes) of carbon dioxide (CO_2) per year and the natural processes can only absorb about half of that amount, so there is a net increase of 10.65 billion tonnes of atmospheric carbon dioxide (USDoE, 2007). Coal combustion also leads to sulphur dioxide (SO_2) emissions with serious implications for local pollution (Shukla, 1997). Biomass burning is also recognized as a significant global source of emissions contributing as much as 40% of gross carbon dioxide and 38% of tropospheric ozone (Levine, 1991). Besides, 1.4 million tonnes of methane (CH_4) emissions are also reported from burning traditional biomass fuels. Apart from these emissions, there are a number of other environmental problems associated with energy use. Thus, the energy system is turning out to be 'unsustainable' in the 21st century.

In recent years, researchers have recognized the importance of holistic thinking. Current Kyoto-based approaches to reduce the earth's greenhouse gas involve seeking ways to reduce emissions. Biofuels have emerged as one of the most strategically important alternative fuel sources and are considered as an important way of progress for limiting greenhouse gas emissions, improving air quality and finding new energetic resources (Delfort et al., 2008). A fuel is considered as biofuel if it is derived from biomass such as agricultural products or residues, industrial and urban residues, wood residuals and forest products, either as liquid or as gas (Granda et al., 2007; Prasad et al., 2007a). It encompasses mainly bioethanol, biodiesel, biogas and biohydrogen (NREL, 2006). Ideally a biofuel should be carbon neutral and should therefore not contribute to the overall accumulation of carbon in the atmosphere (Oliveira et al., 2005). Carbon in crops is the result of the photosynthetic conversion of carbon dioxide in the atmosphere (capturing CO_2) into dry matter determined by solar radiation during the growing season (Tilman et al., 2006) and by natural resources (e.g. climate, water) and external inputs (e.g. fertilizers, pesticides). Biofuel is thus considered an important component of the global strategy to reduce green house gas emission, improve air quality and to increase energy security by providing an alternative to fossil fuels (Farrell et al., 2006; Larson, 2006; Prasad et al., 2007b).

The worldwide investment in new biofuels production capacity has also been growing rapidly and was expected to exceed $4 billion in 2007. The value of biofuels production plants under construction and announced construction plans through 2009 exceeded $4 billion in the United States, $4 billion in Brazil and $2 billion in France (REN21, 2008). Biofuels production technologies, despite their techno economic potential have found meagre deployment due to myriad barriers. Recent developments in global climate change negotiations which culminated in the Kyoto Protocol are likely to remove some of the vital barriers to RETs which permitted fossil fuels to externalize the environmental costs. If biofuels want to be part of the solution they must accept a degree of scrutiny unprecedented in the development of a new industry. That is because sustainability deals explicitly with the role of biofuels in ensuring the well-being of our planet, our economy, and our society both today and in the future (Sheehan, 2009).

There are three key arguments for the commercial use of biofuels:

a. Economic-driven rise in consumption, resulting in higher prices for fossil fuels;

b. Energy security and geo-political dependence of regions with a high volatility;

c. Anthropogenic-based CO_2 emissions and climate change.

2. GLOBAL BIOFUEL SCENARIO

Soaring prices of fossil-fuels and environmental pollution associated with their use has resulted in increased worldwide interest in the production and use of biofuels. Both developed and developing countries have made mix of policies which have triggered public and private investments in biofuel crop research and development and biofuels production (EPA, 2009; REN21, 2008). Biofuels already constitute the major source of energy for over half of the world's population, accounting for more than 90% of the energy consumption in poor developing countries (FAO, 2005). Presently, biofuels production is expanding, especially in Brazil, the USA and South-East Asia, where sugar cane, maize and palm oil are converted into ethanol or biodiesel (Anonymous, 2008). Over the next several decades, the most certain increase in demand for biofuels is going to focus on displacing liquid fuels for transport, mostly in the form of ethanol which currently supplies over 95% of the biofuels for transportation (Fulton et al., 2004). The world's top ethanol fuel producers in 2009 were the United States with 10.75 billion US liquid gallons (bg) and Brazil (6.58 bg), accounting for 89% of world production of 19.53 billion US gallons (73.9 billion liters or 58.3 million metric tonnes according to data assembled by F.O. Licht. The Global Renewable Fuels Alliance (GRFA) is an international federation representing more than 65 per cent of the world's renewable fuels production from 30 countries. The GRFA predicts global production will reach 85.9 billion litres in 2010 growing by 16.2 per cent from 2009 production (Enagri, 2010). Global production of biodiesel has grown rapidly as well, although starting from a much smaller base. Biodiesel output expanded from 0.23 billion gallons in 2000 to 3.9 billion gallons in 2008 (EPA, 2009). The European Union produces nearly 80 percent of the world's biodiesel, largely from rapeseed; Germany is the single largest biodiesel producer, followed by the United States which produces the fuel mainly from soybeans (Nicholas, 2007). According to Pitkanen et al. (2003) lignocellulosic materials could support the sustainable production of liquid transportation fuels. The 73.9 Tg dry wasted crop material worldwide could potentially produce 49.1 GL $year^{-1}$ of bioethanol (Kim & Dale, 2004), about 16 times higher than the current world ethanol production. The potential bioethanol production could replace 353 GL of gasoline i.e. 32% of the global gasoline consumption (Prasad et al., 2007a).

2.1. Biofuel policies in different countries

Policy choices are instrumental in determining the direction of national as well as global biofuels development. Around the world, governments are considering a number of biofuel policy options. The biofuel policy aims to promote the use in transport of fuels made from biomass as well as other renewable fuels. The central policy of biofuel concerns job creation, greater efficiency in the general business environment and protection of the environment. A range of policies are currently being implemented to promote renewable bioenergy in United States, including the Energy Policy Act of 2005, the Energy Independence and Security Act of 2007, the 2002 Farm Bill and the Biomass Research and Development Act of 2000 (FAO, 2008).

Policy targets for renewable energy exist in at least 66 countries worldwide, including all 27 European Union countries, 29 U.S. states (and D.C.) and 9 Canadian provinces. Most targets aim for the 2010–2012 timeframe, although an increasing number of targets aim for 2020. There is now an EU-wide target of 20 percent and a Chinese target of 15 percent of primary energy by 2020. Besides China, several other developing countries adopted or upgraded targets during 2006/2007 (REN21, 2008). China finalized targets for the equivalent of 13 billion liters of ethanol and 2.3 billion liters of biodiesel per year by 2020. The directive sets a European target of 2% substitution of conventional transport fuels by biofuels by December 2005 and a further 5.75% substitution by December 2010. Moreover, the European Commission is committed to encourage the production and use of biofuels by proposing to set a binding minimum target for renewable energy sources of 10% of final energy use in the transport sector by 2020 and is also working on changing fuel specifications to allow higher than 5% blends of biofuel (Lechon, 2009; FAO, 2008).

New U.S. renewable fuels standard requires fuel distributors to increase the annual volume of biofuels blended to 36 billion gallons (136 billion liters) by 2022. The new standard implies that 20 percent of gasoline for road transport would be biofuels by 2022. Several states within the U.S. have also taken steps to promote development and increased use of biofuels. Under the Energy Policy Act of 2005, U.S. renewable transportation fuels are scheduled to reach 7.5 billion gallons by 2012. The 2007 Energy Independence and Security Act require 36 billion gallons of ethanol by 2020, with 21 billion gallons coming from advanced biofuels such as cellulose-based ethanol (Hoekman, 2009). The United Kingdom has a similar renewable fuels obligation, targeting 5 percent by 2010. Japan's new strategy for long-term ethanol production targets 6 billion liters per year by 2030, representing 5 percent of transport energy (FAO, 2008).

Developing countries like India also started Biofuel mission in 2003 to cope with the global fuel crisis. Government of India through a notification in September 2002 made 5% ethanol blending mandatory in petrol in 9 states and 3 Union Territories. In the next phase, supply of ethanol-blended petrol would be extended to the whole country and efforts would be made to increase the percentage of ethanol mixture in petrol to 10 percent (Prasad et al., 2007b). National Biofuel Policy drafted by the Ministry of New and Renewable Energy Sources (MNRE), assures that biofuel programme would not compete with food security and the fertile farm lands would not be diverted for plantation of biofuel crops. The policy deals with a number of issues like minimum support prices (MSPs) for biofuel crops, subsidies for growers of biofuel crops, marketing of oil-bearing seeds, subsidies and fiscal concessions for the biofuel industry, R&D, mandatory blending of auto-fuel with biofuel, quality norms, testing and certification of biofuels. An indicative target of 20% by 2017 for the blending of biofuels–bioethanol and biodiesel has been proposed in the National Biofuel Policy (Indian Express, 2008).

3. BIOMASS RESOURCE BASE AND BIOFUEL GENERATION TECHNOLOGY

Since the mid-1970s many research initiatives have focused on increasing the biomass resource base for production of biofuel. Several technologies used for the conversion of plant material into biofuels are available and depend on the type of feedstock; the conventional and new technologies can be classified into the following four groups:

3.1. First generation biofuel technology

In general, first generation biofuels are produced from cereal crops (e.g. wheat, maize), oil crops (e.g. rape, palm oil) and sugar crops (e.g. sugar beet, sugar cane) using established technology. Based on the conversion of sugars (sugar cane) and starch (potato, cassava, maize) or oil (oil palm, rapeseed) accumulated in food crops into ethanol and biodiesel respectively accounts the first generation biofuels (Cassman & Liska, 2007). Some have called for an integrated systems biology approach to define ideotypes that meet the requirements of feedstocks for biofuels. However, the scientific evidence that crop traits can be genetically modified to meet the requirements for fuel without any trade-off on the value as a food crop is absent. Alternatively, different varieties may be developed for food and fuel production.

3.2. Second generation biofuel technology

In general, second generation biofuels are produced from cellulosic materials (Somerville, 2006) and also based on the use of dedicated energy crops like switch grass *(Panicum virgatum)* grown with low external inputs and using conversion methods that result in high net energy efficiency (output/input). Conversion of cellulosic biomass, which is both abundant and renewable, is considered as a promising alternative for ethanol produced from starch or sugar. Plant triacylglycerols are another potential feedstock to produce biofuels, especially biodiesel. Most vegetable oils are derived from triacylglycerols stored in seeds. Novel energy crops may be developed that produce triacylglycerols in non-seed tissues (Durrett et al., 2008). To avoid competition with food crops there is a growing interest in woody/tree borne oil plants. Native energy oil plants are more frequently present in tropical and subtropical regions. Non-edible oils obtained from plant species such as *Jatropha curcas* (Ratanjyot), *Pongamia pinnata* (Karanj), *Calophyllum inophyllum* (Nagchampa), *Hevea brasiliensis* (Rubber) and other oil-based crops can be efficiently used for biodiesel production. *Jatropha curcas* is a drought resistant, perennial oil plant (ca. 40% oil content) with favourable traits to produce biodiesel in unfavourable regions of India, Sub-Sahelian Africa and Latin America (Kumar & Sharma, 2008).

3.3. Third generation biofuel technology

This is based on algae or cyanobacteria that contain a high oil mass fraction (up to 70%) and are grown in ponds. Microalgae are sunlight-driven cell factories that convert carbon dioxide to potential biofuels (Akkerman et al, 2002; Ghirardi et al., 2000). Oil content in microalgae can exceed 80% by weight of dry biomass (Metting, 1996). Depending on species, microalgae produce many different kinds of lipids, hydrocarbons and other complex oils (Guschina & Hardwood, 2006; Metzger & Largeau, 2005). Certain algae and cyanobacteria have high lipid contents (Spolaore et al., 2006) biodiesel derived from microalgal oil (Dunahay et al., 1996; Sheehan et al., 1998). Under proper conditions, these micro-organisms can produce lipids for biodiesel with yields per unit area that are 50-100% higher than those with any plant system (Chisti, 2008). Microalgae can also provide several different types of renewable biofuels which include methane produced by anaerobic digestion of the algal biomass (Gavrilescu & Chisti, 2005) and photobiologically produced biohydrogen (Fedorov et al., 2005; Kapdan & Kargi, 2006).

The microalgae appear to be the only source of renewable biodiesel that is capable of meeting the global demand for transport fuels which will not compromise production of food, fodder and other products derived from crops (Chisti, 2008; Hu et al., 2008). However, it is still not proven that this high efficiency can be maintained after scaling-up the technology to a large production plant. Furthermore, the feedstock is waste derived from plant material used for food and feed. Yet, we do not know what the trade-off is between maximizing the utilization of primary production for food and feed and the use of residues and waste to produce methane or hydrogen. Chemical composition of the residues and waste also will matter.

3.4. Fourth generation biofuel technology

It is based on biohydrogen production by embedding parts of the photosynthesis apparatus in artificial membranes (Kruse et al., 2005). These biofuels are derived from the bioconversion of living organisms (microorganisms and plants) using biotechnological tools. The mean conversion efficiency for the total solar spectrum amounts to ca. 20%, which is on average about 10 times higher than for annual crops. This high efficiency should be considered a potential level. The gap between the potential level and actual efficiency is still not known. Currently, this technology is still expensive and not yet ready for commercial exploitation.

4. BIOFUELS FOR IMPROVEMENT IN AIR QUALITY

The emissions from engines using gasoline have SO_2, CO_2, VOCs, nitrogen oxides (NOx), carbon monoxide (CO) and particulate matter (PM), which causes pollution (Gaffney & Marley, 2009). The VOCs emitted from gasoline-fueled vehicles arise from uncombusted or partially combusted fuel and typically include cyclohexane, octanes, and aromatics. The NOx and VOC emissions react in the presence of sunlight by way of a series of photochemical reactions involving hydroxyl, peroxy and alkoxy radicals, to form the secondary pollutant ozone (Finlayson-Pitts & Pitts, 2000). The emissions of NOx and total VOCs lead to the formation of ozone in the troposphere, the main component of smog. CO is a deadly poison and the inhalation of fine particulate matter (PM2.5) is a serious health concern (Peter et al., 2003).

The energy and environmental crisis which the world is experiencing forced to find alternative uses for renewable resources and using clean technologies. Feedstocks include agricultural and food processing wastes, trees, and various grasses that are converted to ultra-clean (minimal SOx and NOx pollutants) biofuel in elaborate biochemical or thermo-chemical steps. And depending on the choice of a microorganism the bio-conversion can yield cellulosic ethanol, biogas or biohydrogen. Biofuels has a number of health and environmental benefits including improvement in air quality by reducing pollutant gas emissions relative to fossil fuels (Vasudevan et al., 2005). Therefore, it is imperative to develop and promote alternative energy sources that can lead to sustainability of the energy system. Hall & House (1993) have examined the role of biomass in mitigating global warming and contributing to the development of future energy strategies and concluded that the use of biomass for fossil fuel substitution would be far more effective in reducing atmospheric CO than to simply sequester CO_2 in forests in most circumstances. Currently, the second generation biofuels are projected to reduce carbon emissions by 90%, and by 2040 these could potentially replace up to 40% of all conventional fuels (Krisztina et al., 2010).

4.1. Combustion profile of biofuels

The success of oxygenated gasoline has sparked interest in the use of oxygenated compounds as emissions reducing additives in diesel fuel. Oxygenated compounds used as diesel additives are structurally similar to diesel fuel but have one or more oxygen atoms bonded to the hydrocarbon chain. Numerous oxygenated compounds have been investigated as either diesel fuel additives or replacements and have shown emissions reducing properties.

4.1.1. Properties and combustion profile of ethanol

Although ethanol was always a good oxygenate candidate for gasoline, the compound first approved by Environmental Protection Agency was methyl tertiary butyl ether (MTBE), a petrochemical industry product (Gaffney &

Marley, 2000). The introduction of MTBE in gasoline has been studied as a classic case of solving one problem (reducing vehicle carbon monoxide emissions) while causing a new problem (persistent contamination of water systems with MTBE). Use of MTBE increased until 1999, but reports then appeared of environmental pollution incidents caused by MTBE spillage; US bans on MTBE came into force during 2002. Presently, ethanol is prospective material for use in automobiles as an alternative to petroleum based fuels. The main reason for advocating ethanol is that it can be manufactured from natural products or waste materials, compared with gasoline, which is produced from non-renewable natural resources. Ethanol can be independently used as a transportation fuel together with additives (e.g. ignition improver, denaturing agents, etc.). In addition, instead of pure ethanol, a blend of ethanol and gasoline is a more attractive fuel with good anti-knock characteristics (Al-Hasan, 2003).

Ethanol contains 34.7% oxygen by weight, and adding oxygen to fuel results in more complete fuel combustion, and therefore contributes to a reduction in exhaust emission and petroleum use (Huang et al., 2008; Prasad et al., 2007b). Ethanol is a high octane fuel and its use displaces toxic octane boosters such as benzene, a carcinogen. Ethanol is a virtually sulfur free additive and is biodegradable. Thus, it's easy to see why many states use ethanol to reduce vehicular emissions. The physical and thermo-physical properties of ethanol compared to the other fuels (gasoline and diesel) indicates that ethanol is more suitable and environmentally safe fuel (Table 1) as its normal boiling point lies in between gasoline and diesel, while heating value, carbon and sulfur content are lower (Lynd, et al., 1991; Vaivads et al., 1995).

Table 1. Comparison of thermo-physical properties of ethanol, gasoline and diesel fuel

Properties	Ethanol	Gasoline	Diesel
Density (g cm^{-3})	0.785	0.737	0.856
Normal boiling point (°C)	78.00	38-204	125-400
Lower heating value, LHV (kJ cm^{-3})	21.09	32.05	35.66
LHV (kJ g^{-1})	26.87	43.47	41.66
Energy (MJ l^{-1})	23.10	32.84	33.32
Energy (MJ kg^{-1})	29.40	47.46	46.94
Carbon content (%)	52.20	85.50	87.00
Sulfur content (ppm)	0.00	~200	~250

A comparison of flammability variables for neat diesel, ethanol and gasoline clearly showed that ethanol (Table 2) falls between diesel and gasoline in terms of flashpoint and flammability temperature limits (Battelle, 1998). In the engine durability tests conducted by Meiring and coworkers (1983), no abnormal deterioration of the engine or fuel injection system was detected after 1000 hrs of operation on a blend containing 30% dry ethanol, small amount of octyl nitrate ignition improver and ethyl acetate phase separation inhibitor and the remainder diesel fuel. The Chicago Transit Authority in the US monitored the condition and overall performance of a fleet of 30 buses, of which 15 were the control run on number one diesel. After completion of 434,500 km distance by the 15 buses running on the blend, no abnormal maintenance or fuel related problems were encountered (Marek & Evanoff, 2001).

Table 2. Approximate fuel ethanol characteristics related to flammability

Characteristics	Neat diesel	Neat ethanol	Neat gasoline
Vapour-pressure at 37.8 °C (kPa)	0.3	17	65
Flash point (°C)	64	13	-40
Auto-ignition temperature (°C)	230	366	300
Flammability limits (%)	0.6-5.6	3.3-19.0	1.4-7.6
Flammability limits (°C)	64-150	13-42	-40-18

Low-percentage ethanol-gasoline blends (5-10%) can be used in conventional spark-ignition engines with almost no technical change. New flex-fuel vehicles of which there are over 6 million running mainly in Brazil, United States and Sweden, can run on up to 85% ethanol blends that had modest changes made during production. Ethanol combustion offers fuel and emissions savings due to the high octane number, the high compression ratio and the combustion benefits from ethanol vapour cooling which partly offsets its lower energy content per liter (IEA-ETE, 2007).

4.1.2. Properties and combustion profile of biodiesel

Biodiesel is a mono-alkyl ester based oxygenated fuel made from vegetable oil or animal fats. It has properties similar to petroleum based diesel fuel and can be blended into conventional diesel fuel. This interest is based on a number of properties of biodiesel, non toxic and its potential to reduce exhaust emissions (Jha, 2009; Knothe et al., 2006). The advantages of biodiesel as diesel fuel are its portability, ready availability, renewability, higher combustion efficiency, lower sulfur and aromatic content (Knothe et al., 2006; Ma & Hanna, 1999), higher cetane number, and higher biodegradability (Mudge & Pereira, 1999; Speidel et al., 2000; Zhang et al., 2003). Biodiesel is by nature is an oxygenated fuel with oxygen content of about 10%. This improves combustion and reduces CO, soot and unburnt hydrocarbon.

Biodiesel is non-flammable and, in contrast to petrodiesel, is non explosive. The flash point of biodiesel (>130 °C) is significantly higher than that of petroleum diesel (64 °C) or gasoline (−45 °C) (Anonymous, 2010a). Biodiesel has a density of ~0.88 g/cm^3, higher than petrodiesel (~0.85 g/cm^3). Biodiesel has better lubricating properties and much higher cetane ratings than today's lower sulfur diesel fuels (Knothe et al., 2005; Mittelbach & Remschmidt, 2004). Biodiesel addition reduces fuel system wear (Anonymous, 2010b) and in low levels in high pressure systems increases the life of the fuel injection equipment that relies on the fuel for its lubrication. The calorific value of biodiesel is about 37.27 MJ/L (Elsayed et al., 2003). Variations in biodiesel energy density are more dependent on the feedstock used than the production process and properties of biodiesel from different oils are shown in Table 3 (Chhang et al., 1996; Rao & Gopalakrishnan, 1991). Biodiesel has virtually no sulfur content, and it is often used as an additive to Ultra low sulphur diesel (ULSD) fuel to aid with lubrication, as the sulfur compounds in petrodiesel provide much of the lubricity.

Table 3. Approximate fuel biodiesel characteristics related to flammability

Biodiesel from Vegetable oil	Kinematic Viscosity mm^2/s	Cetane no:	Heating value MJ/kg	Flash Point °C	Density kg/l
Peanut	4.9	54	33.6	176	0.883
Soybean	4.5	45	33.5	178	0.885
Babassu	3.6	63	31.8	127	0.875
Palm	5.7	62	33.5	164	0.880
Sunflower	4.6	49	33.5	183	0.860
Diesel	3.06	50	43.8	76	0.855
B20 (20% blend)	3.2	51	43.2	128	0.859

Since the key properties of the biodiesel are comparable to those of diesel fuel, it can be used in all diesel engines with little modification or no modification either on its own or as a blend with conventional or low sulphur diesel (Ryan, 1999). The disadvantages of biodiesel are its higher viscosity, lower energy content, higher cloud point and pour point, higher nitrogen oxide (NOx) emissions, lower engine speed and power, injector coking, engine compatibility, high price and greater engine wear. The technical disadvantages of biodiesel fossil diesel blends include problems with fuel freezing in cold weather, reduced energy density and degradation of fuel under storage for prolonged periods. However there are solutions to this such as using a blend of biodiesel upto B20 which has a gelling point of –15 degrees F, adding a biodiesel additive such as Fuel Boost to the blend also lowers the gel point even further and useful in the winter (Petracek, 2011).

4.1.3. Properties and combustion profile of biogas

Biogas is a renewable fuel produced by anaerobic fermentation of organic material (Pathak et al., 2009). The value of a substrate in the biogas process depends on its potential as a high yield plant species and on the quality of the biogas produced such as the achievable methane content. The most suitable plant species for the production of biogas are those which are rich in degradable carbohydrates such as sugars, lipids and proteins, and poor in hemicelluloses and lignin, which have a low biodegradability (El Bassam, 1998). Its composition varies with the source, but usually it has 50–70% CH_4, 25–50% CO_2, 1–5% H_2, 0.3–3% N_2 and traces of H_2S (Bedoya, 2009). Methane is the only combustible constituent of biogas, which is utilized in different forms of energy. Biogas can be used for heating, lighting, transportation, small-scale power generation, and large gas turbines as a complementary fuel (e.g., to natural gas) (Bedoya, 2009). Constraints like cost of cleaning, upgrading (to remove CO_2) and transportation of biomass limit the use of biogas (Jahangirian et al., 2009).

Methane is very light fuel gas. If we increase the number of hydrogen and carbon atoms, we have got progressively heavier gases, releasing more heat, therefore more energy, when ignited. Specific gravity of methane is 55 which is less than petrol & LPG. This means that biogas will rise if escaping, thus dissipating from the site of a leak. This important characteristic makes biogas safer than other fuels. It does not contain any toxic component; therefore there is no health hazard in handling of fuel. The calorific value of biogas is 5000-7000 Kcal/m^3. In calorific value, one cubic meter of biogas is equivalent to 0.7 m^3 of natural gas, 0.7 kg of fuel oil and 4 kWh of electricity (Asankulova & Obozov, 2007).

Motive power can be generated by using biogas in dual fuel internal combustion (IC) engine. Air mixed with biogas is aspirated into the engine and the mixture is then compressed, raising its temperature to about 350°C, which is the self-ignition temperature of diesel. Biogas has a high (600°C) ignition temperature. Therefore, in order to initiate combustion of the charge, a small quantity of diesel is injected into the cylinder just before the end of compression. The charge is thus ignited and the process is continued smoothly. Converting a spark-ignition engine for biogas fueling requires replacement of the gasoline carburettor with a mixing valve (pressure-controlled venturi type or with throttle). A spark-ignition engine (gasoline engine) draws a mixture of fuel (gasoline or gas) and the required amount of combustion air. The charge is ignited by a spark plug at a comparably low compression ratio of between 8:1 and 12:1. Power control is affected by varying the mixture intake via a throttle (Biogas Digest, 2010). Biogas has very high octane number approximately 130. By comparison, gasoline is 90 to 94 & alcohol 105 at best. This means that a higher compression ratio engine can be used with biogas than petrol. Hence, cylinder head of the engine is faced so that clearance volume will be reduced and compression ratio can sufficiently increase. Thus volumetric efficiency and power output are increased.

4.2. Biofuels for GHGs emission reduction and air quality

Vehicular emissions from petroleum products in the form of CO, NOx, unburnt hydrocarbons and particulates are of high environmental concern especially in air pollution (Subramanian et al., 2005). Thermal power plants are a major source of SPM (suspended particulate matter) and solid waste. The inefficient burning of biomass causes exposure to various pollutants and is considered a major health hazard and has been shown to lead to lung and chest problems among women and children (Smith, 1987). Biofuels has a number of health and environmental benefits including improvement in air quality by reducing pollutant gas emissions relative to fossil fuels (Vasudevan et al., 2005). Therefore, it is imperative to develop and promote alternative energy sources that can lead to sustainability of the energy system. This would not only warrant major reforms in the energy policies and infrastructure, but also huge international investments.

4.2.1. Reduction in exhaust emission by ethanol

Ethanol is one of the best tools available today to reduce air pollution from vehicles. Ethanol-diesel emulsion gives beneficial results in terms of pollution emission reduction in engines (Jha, 2009; Knothe et al., 2006). It is found that a remarkable improvement in PM-NOx trade-off can be achieved by promoting the premixing based on the ethanol blend fuel having low evaporation temperature, large latent heat and low cetane number as well, in addition, based on a marked elongation of ignition delay due to the low cetane number fuel and the low oxygen intake charge (Ishida et al., 2010). As a result, very low levels of NOx and PM which satisfies the 2009 emission standards imposed on heavy duty diesel engines in Japan, were achieved without deterioration of brake thermal efficiency in the PCI engine fuelled with the 50% ethanol blend diesel fuel and the high exhaust gas recirculation (EGR) ratio. It is noticed that smoke can be reduced even by increasing the EGR ratio under the highly premixed condition (Ishida et al., 2010). A 41% reduction in particulate matter and 5% NOx and 27% CO emission has been observed with 15% ethanol blends. Emission tests conducted especially on ethanol-diesel blends (Table 4) confirm the effect of substantially reducing particulate matter (Prasad et al., 2007b).

Table 4. Reduction in pollution emission with different percentages of Ethanol blending

Pollutant	Emission (%)		Emission (g/km)	
	10% ethanol	15 % ethanol	22% ethanol	100 % ethanol
Particulate matter	27	41	0.08	0.02
NOx	4	5	0.45	0.34
Carbon monoxide	20	27	0.76	0.65
Unburned hydrocarbons	-	-	0.004	0.02
Sulfur dioxide	-	-	0.064	0.0

If blended at the refinery, as opposed to "splash blending" outside the refinery, ethanol-blended gasoline can reduce NOx emissions as well, thus further reducing the potential for smog. Compared with conventional unleaded gasoline, ethanol is a particulate-free burning fuel source that combusts with oxygen to form carbon dioxide, water and aldehydes. Gasoline produces 2.44 CO_2 equivalent kg/l and ethanol 1.94 (Popa, 2010). Since ethanol contains 2/3 of the energy per volume as gasoline, ethanol produces 19% more CO_2 than gasoline for the same energy. When compared to gasoline, depending on the production method, ethanol releases less green house gases and savings of GHG emissions from ethanol produced from various crops are seen (Wang et al., 2009). Ethanol could play an important role in reducing petroleum consumption by enabling a substantial increase in the fuel efficiency of gasoline engine vehicles. This ethanol boosted engine concept uses a small amount ethanol to increase the efficiency of use of a much larger amount of gasoline by approximately 30%. Gasoline consumption and the corresponding CO_2 emissions would thereby be reduced by approximately 25%. In combination with the additional reduction that results from the substitution of ethanol for gasoline as a fuel, the overall reduction in gasoline consumption and CO_2 emissions is greater than 30% (Cohn et al., 2005).

4.2.2. Atmospheric pollution reduction by biodiesel

Biodiesel is a clean-burning renewable fuel that is compatible with petroleum diesel and can be produced domestically. The biodiesel performs as well as diesel while reducing the emissions of particulate matter, carbon monoxide (CO), hydrocarbons, oxides of sulphur (SOx), particulate matter and smoke density (Ali et al., 1995; Bagley et al., 1998; Durbin et al., 2000; Koo & Leung, 2000). Biodiesel is considered as 'carbon neutral' because all the carbon dioxide (CO2) released during consumption had been sequestered from the atmosphere for the growth of vegetable oil crops (Barnwal and Sharma, 2005). Other environmental benefits of biodiesel include the fact that it is highly biodegradable and appear to reduce emissions of air toxics and carcinogens (relative to diesel). The benefits of 100% (B 100) and 20% (B 20) biodiesel blending, in terms of per cent pollutants emission reduction (Planning Commission of India, 2003) and reduction emission in g/km for 10 and 15 % blend (Vasudevan et al., 2005) is shown in Table 5. According to the EPA's Renewable Fuel Standards Program Regulatory Impact Analysis, released in February 2010, biodiesel from soy oil results an average of 57% reduction in greenhouse gases compared to fossil diesel, and biodiesel produced from waste grease results in an 86% reduction (Petracek, 2011).

Table 5. Reduction in pollution emission with different percentages of biodiesel blending. *(-) and (+): Less and more % of pollutant emission from biodiesel in comparison to 100% diesel

Pollutant	Emissions reduction (%)		Emission (g/km)		
	B 100	B20	Diesel	B 10	B 15
Particulate matter	-30	-22	0.129	0.093	0.080
NOx	+13	+2	0.79	0.83	0.89
Carbon monoxide	-50	-20	0.77	0.65	0.62
Unburned hydrocarbons	-93	-30	0.37	0.22	0.16
Sulfur dioxide	-100	-20	--	---	---

*(-) and (+): Less and more % of pollutant emission from biodiesel in comparison to 100% diesel

Biodiesel has higher cetane number, lower sulfur content and lower aromatics than that of conventional diesel fuel. It also reduces emissions due to presence of oxygen in the fuel (Subramanian et al., 2005). In addition, the exhaust emissions of sulfur oxides and sulfates (major components of acid rain) from biodiesel are essentially eliminated compared to diesel. Of the major exhaust pollutants, both unburned hydrocarbons and nitrogen oxides are ozone or smog forming precursors.

The use of biodiesel results in a substantial reduction of unburned hydrocarbons. However, a marginal increase in NOx (1-6%) is reported (Table 5) for biodiesel use in many engines. Emissions of nitrogen oxides are either slightly reduced or slightly increased depending on the duty cycle of the engine and testing methods used. Based on engine testing, using the most stringent emissions testing protocols required by EPA for certification of fuels or fuel additives in the U.S., the overall ozone (smog) forming potential of the hydrocarbon exhaust emissions from biodiesel is nearly 50 percent less than that measured for diesel fuel (Petracek, 2011). The summary report given by NREL stated that the maximum estimated increase and decrease in daily maximum 1-hour or 8-hour ozone concentrations due to the use of either a 100% or 50% penetration of a B20 fuel in the HDDV fleet in any of the areas studied is +0.26 ppb and –1.20 ppb for 1-hour ozone and the 100% B20 fuel scenario. As the maximum ozone increase (+0.26 ppb) is well below 1 ppb, the use of biodiesel is estimated to have no measurable adverse impact on 1-hour or 8-hour ozone attainment in Southern California and the Eastern United States (Morris et al., 2003). The mass concentration of the particles/smoke decreased up to 33% when the engine burned 100% biodiesel as fuel, compared to the 100% petroleum diesel (Zou and Atkinson, 2003).

4.2.3. Atmospheric pollution reduction by biogas

The fossil fuels combustion leads to emission of air pollutants such as CO, NOx, SO_2, volatile organic compounds and particulates (Parashar et al., 2005). Biogas technology, besides supplying energy and manure, provides an excellent opportunity for reducing environmental hazards and pollution through substituting firewood for cooking, kerosene for lighting and cooking and chemical fertilizers (Pathak et al., 2009). The benefits of biogas are generally similar to those of natural gas. In addition, burning biogas reduces greenhouse gas (GHG) emissions; it reduces the net CO_2 release and prevents CH_4 release. Thus, biogas combustion is a potential means to satisfy various legislative and ecological constraints (Jahangirian et al., 2009). Borjesson & Berglund (2006) analyzed fuel-cycle emissions of CO2, CO, NOx, SO_2, hydrocarbons (HC), CH_4, and particles from a life-cycle perspective for biogas systems based on different digestion technologies and raw materials. They suggest that the overall environmental impact of biogas depends largely on the status of uncontrolled losses of CH_4, the end-use technology that is used, the raw material digested, and the energy efficiency in the biogas production chain.

Biogas is a smokeless fuel offering an excellent substitute for kerosene oil, cattle dung cake, agricultural residues and firewood which are used as fuel in most of the developing countries (MNES, 2006). Burning of kerosene, firewood and cattle dung cake as fuels emits 0.8 to 2.2, 0.7 to 4.0 g kg^{-1} NOx, and SO_2, respectively along with varying amounts of CO, volatile organic compounds, particulate matters, organic matter, black carbon and organic carbon (Table 6).

A family size biogas plant substitutes 316 L of kerosene, 5,535 kg firewood and 4,400 kg cattle dung cake per annum as fuels. Substitution of kerosene reduces emissions of NOx, SO_2 and CO by 0.7, 1.3, and 0.6 kg $year^{-1}$. Substitutions of firewood and cattle dung cake results in the reduction of 3.5 to 12.2, 3.9 to 6.2, 436.9 to 549.6 and 30.8 to 38.7 kg $year^{-1}$ NOx, SO_2, CO and volatile organic compounds, respectively. Total reductions of NOx, SO_2, CO and volatile organic compounds by a family size biogas plant are 16.4, 11.3, 987.0 and 69.7 kg $year^{-1}$ (Pathak et al., 2009).

Table 6. Pollution reductions due to use of biogas plant

Pollutants	Pollution reduction due to a biogas plant (kg $year^{-1}$)			
	Kerosene	Firewood	Dung cake	Total
Oxides of N (NOx)	0.7	12.2	3.5	16.4
Oxides of S (SO_x)	1.3	3.9	6.2	11.3
Carbon monoxide	0.6	549.6	436.9	987.1
Volatile organic compounds	0.2	38.7	30.8	69.7
Particulate matter^{10}	0.1	16.6	13.2	29.9
Particulate matter$^{<2.5}$	0.1	11.6	28.6	40.3
Organic matter	0.4	7.2	17.6	25.2
Black carbon	0.1	3.3	11.0	14.4
Organic carbon	0.1	19.4	55.4	74.9

The biogas used as vehicle fuel presents better characteristics than the natural gas (Table 7). Some disturbance still appears for the NOx emissions, but they stay below the EU norms. Concerning CO_2, hydrocarbons and CO emissions, the biogas is far better than the Natural Gas used for Vehicles (NGV), (Traffic & Public Transport Authority, 2000).

Methane has a greenhouse gas (GHG) heating factor 21 times higher than CO_2. Combustion of biogas converts methane into CO_2 and thereby reduces the GHG impact by over 20 times.

Table 7. Pollution reductions due to biogas used as vehicle fuel

Pollutant	Emission (g/km)		
	Diesel	Natural Gas	Biogas
Particulate matter	0.1	0.022	0.015
NOx	9.73	1.1	5.44
Carbon monoxide (CO)	0.2	0.4	0.08
Unburned Hydrocarbons (HC)	0.4	0.6	0.35
CO2	1053	524	223

Combustion of biogas reduces the flame temperature, which reduces NOx emissions since the main pathway for NOx formation is thermal (Lafay et al., 2007). The digester reduces emissions of methane, carbon dioxide and ammonia from manure while in the enclosed vessel. Combustion of the biogas releases some carbon dioxide and sulphur compounds back into the atmosphere. However this combustion process releases carbon dioxide, which was captured by plants in the last year by the crop fed to the animals in contrast to fossil fuels, which are releasing carbon from ancient biomass.

4.3. Effect of biofuels on health

The exhaust gases from transportation vehicles contain many types of gaseous and particulate air pollutants, including trace levels of some particulate polycyclic aromatic hydrocarbons (PAHs) which have adverse effects on human health (Prasad et al., 2007b; Subramanian et al., 2005). Burning of biomass or any solid fuel, most closely associated with air quality problems and has some negative impacts on health (Pathak et al., 2009), particularly when burned in household cooking/heating stoves where there is little or no ventilation. Exposure to particulates from biomass burning causes respiratory infections in children, and carbon monoxide is implicated with problems in pregnancy. Coal and biomass are also suspected of causing cancer, where exposure rates are high (Smith, 1993). Petroleum fuels produce aromatic compounds of a polycyclic nature which are responsible for producing cancer in humans. But increased levels of NOx and HC may effects the human health as these may contain carcinogenic HC as well. If these productions can be reduced then considerable reduction in cancer amongst human beings can be hoped for. So for all of these reasons and biofuel production should be increased to improve our environmental as well as physical health (Wang et al., 1997).

It is highly likely that the net public health impact of using biofuels is beneficial. This is likely true even if the alleged negative impacts of ethanol and biodiesel blending (NOx, permeation) are assumed to be true. This theory is supported by the fact that: (1) ethanol and biodiesel blending significantly reduces emissions of pollutants that are generally believed to pose the greatest public health threat (PM and Toxics i.e. Hazardous Air Pollutants or HAPs); and (2) the actual ozone impact of the alleged increases in NOx and permeation emissions, if assumed to be true, is negligible or extremely small (Coleman, 2011). Ozone levels are significantly increased, thereby increasing photochemical smog and aggravating medical problems such as asthma (Hulsey, 2006; Jacobson, 2007).

4.3.1. Bioethanol and human health

On the positive side, the use of alcohols and alcohol/petroleum blends in diesel engines has been shown to reduce emissions of the potentially carcinogenic carbonaceous soot particles (Gaffney et al., 1980; Wang et al., 1997). Dynamometer studies of the use of gasahol (10% ethanol in gasoline) in motor vehicles report an average decrease in total HC emissions of 5%, a decrease in CO emissions of 13% with an increase in NOx emissions of 5% (HEI, 1996). The same studies showed a decrease in the emissions of the air toxics, benzene and 1, 3-butadiene of 12% and 6%, while acetaldehyde emissions increased by 159%. Although the atmospheric reactivity of ethanol is much lower than that of gasoline, no significant change was reported in the overall atmospheric reactivity (Maximum Individual Risk, MIR) of the exhaust emissions from gasohol when the higher reactivity of acetaldehyde is included. In terms of the health-related PAH emissions, some marked reductions were demonstrated for less toxic gaseous PAHs such as naphthalene, but the particulate PAH emissions, which have more implications for adverse health effects, remaining virtually unchanged and did not show a statistically significant reduction (Zou & Atkinson, 2003).

4.3.2. Biodiesel and human health

The use of biodiesel in a conventional diesel engine results in a substantial reduction of unburned HC, CO and particulate matter compared to emissions from diesel fuel (Table 5). Biodiesel exhaust emission has been extensively characterized under field and laboratory conditions. Biodiesel reduces emissions of CO and CO_2 on a net lifecycle basis and contain fewer aromatic hydrocarbons. Biodiesel can also reduce the tailpipe emission of particulate matters. Vellguth (1983) proved that rapeseed oil methyl esters (RME) are an adequate substitute for fossil diesel fuel (DF). Bünger and his coworkers (1998) investigated the mutagenic and cytotoxic effects of diesel engine exhaust (DEE) from a modern passenger car using rapeseed oil methyl esters (RME) biodiesel as fuel and directly compared to DEE of DF derived from petroleum. The results indicated a higher mutagenic potency of DEE of DF compared to RME due to the lower content of polycyclic aromatic compounds (PAC) in RME exhaust. The existing engines can use 20% biodiesel blend without any modification and reduction in torque output (Vasudevan et al., 2005). The use of a B20 fuel in the HDDV fleet is estimated to reduce the per million risk of premature death due to exposure to air toxics in the SoCAB region of southern California by approximately 2% and 5% respectively (Table 8) for the 50% and 100% HDDV fleet penetration of B20 biodiesel in the HDDV fleet emission scenarios calculated with no indoor/outdoor (I/O) effects and accounting for I/O effects on an annual average and hourly basis, (Morris et al., 2003).

Table 8. Average risk (out of a million) of premature death for the standard diesel base case and the 50% and 100% penetration of B20 biodiesel in the HDDV fleet emission scenarios

Scenario	Std Diesel	50% B20 Fuel		100% B20 Fuel	
	Risk	Risk	(%)	Risk	(%)
No I/O Effects	1950	1910	-2.1	1835	-5.9
Annual I/O Effects	1284	1261	-1.8	1216	-5.3
Hourly I/O Effects	1257	1235	-1.8	1191	-5.3

Scientific research confirms that biodiesel exhaust has a less harmful impact on human health than petroleum diesel fuel. Pure biodiesel emissions have decreased levels of polycyclic aromatic hydrocarbons (PAH) and nitrited PAH compounds that have been identified as potential cancer causing compounds. Also, particulate matter, an emission linked to asthma and other diseases, is reduced by about 47 percent, and carbon monoxide, a poisonous gas, is reduced by about 48 percent (Sinobioenergy, 2011). Biodiesel is the only alternative fuel to have fully completed the health effects testing requirements of the 1990 Clean Air Act Amendments as biodiesel produces less sulfur emissions than regular diesel. The public health benefits of reduced particulate and HAP exposure from biofuels outweigh the negligible smog impact of any relative small NOx and permeation emissions increases from biofuels blends (Coleman, 2011).

4.3.3. Biogas and health benefits

Biogas can have significant health benefits especially in rural areas. According to the Integrated Environmental Impact Analysis carried out by Biogas Support Program for 600 biogas users and 600 non-users, four percent more non-biogas users have respiratory diseases (Tables 9) than those who own biogas plants (BSP, 2000).

Table 9. Health benefits of biogas

Disease	Problems in the past (HHs)*		Present status of HHs	
	Yes	No	Improved	Remained same
Eye Infection	72	18	69	3
Cases of burning	29	71	28	1
Lung problem	38	62	33	5
Respiratory problems	42	58	34	8
Asthma	11	89	9	2
Dizziness/headache	27	93	16	11
Intestinal/diarrhea	58	42	14	44

Qualitative information from various household surveys carried out by BSP has revealed that problems like respiratory illness, eye infection, asthma and lung problems have decreased after installing a biogas plant. According to the Biogas Users' Survey conducted in 2000 with 100 households (HHs*), biogas can have positive impacts on the health of its users. Out of 42 respondents who had respiratory problems in the past, it was reported that the problem has improved for 34 of them. Similarly, those who had problems like asthma, eye infections and lung problems found that their problems had decreased after displacing dirtier fuels with biogas. If parasitic diseases had previously been common, the improvement in hygiene also has economic benefits (reduced working time). The more fully the sludge is digested, the more pathogens are killed. High temperatures and long retention times are more hygienic. The following are the principal organisms killed in biogas plants: Typhoid, Paratyphoid, Cholera and dysentery bacteria (in one or two weeks), Hookworm and bilharzia (in three weeks), Tapeworm and roundworm die completely when the fermented slurry is dried in the sun. Biogas has a positive effect more on rural health conditions.

5. FUEL ECONOMY IN BIOFUEL BLENDS ENGINES

Ethanol (E100) consumption in an engine is approximately 51% higher than for gasoline since the energy per unit volume of ethanol is 34% lower than for gasoline (Chauhan et al., 2011). The higher compression ratios in an ethanol-only engine allow for increased power output and better fuel economy could be obtained with lower compression ratios than gasoline-powered engines. In flexible fuel vehicles, the lower compression ratio requires tunings that give the same output when using either gasoline or hydrated ethanol. A 2004 MIT study (Stauffe, 2006), and an earlier paper published by the Society of Automotive Engineers, identified a method to exploit the characteristics of fuel ethanol substantially better than mixing it with gasoline (Stokes et al., 2000). The improvement consists of using dual-fuel direct-injection of pure alcohol (or the azeotrope or E85) and gasoline, in any ratio up to 100% of either, in a turbocharged, high compression-ratio, small-displacement engine having performance similar to an engine having twice the displacement. Direct cylinder injection raises the already high octane rating of ethanol up to an effective 130 and resulted in over-all reduction of gasoline use and CO_2 emission of 30%.

Biodiesel blends can reduce emission levels of HC (hydrocarbons) and CO (carbon monoxide); however, biodiesel blends may somewhat increase emission levels of NOx (oxides of nitrogen) in some engines. Biodiesel blends, used in new, low emissions engines may not significantly affect emissions. B20 is most widely used by fleets in the United States, because B20 balances performance, EPA emission levels, costs, and availability. B20 is also the minimum blend level that qualifies as an alternative fuel, in compliance with the Energy Policy act of 1992. Blends lower than B20 are used regionally, depending on favorable tax incentives that vary from state to state. However, NOx and evaporative VOCs (permeation) are regulated to control ozone formation, and recent air shed model runs suggest that the use of ethanol (E10) and biodiesel (up to B20) do not measurably increase actual ozone levels. With regard to permeation emissions, it is useful to remember that permeation (an evaporative VOC) is a very small percentage of any state's overall gasoline hydrocarbon emissions inventory (e.g. ~ 4% in California), and that ethanol generally reduces tailpipe (i.e. non-evaporative) hydrocarbon emissions as (at least a partial) offset. Also, because ambient temperature is a primary catalyst for fuel permeation, states with colder climates than California will have much lower permeation rates (Coleman, 2011).

6. FURTHER SCOPE OF BIOFUELS ON ENVIRONMENTAL BENEFITS

Tackling air pollutions and climate change requires the simultaneous deployment of available commercial clean technologies, demonstration and commercialisation of technologies at the advanced research, development and demonstration stage and research into new technologies. So for centuries, biofuels has been playing a vital role in the provision of energy services at the household level. However, at the beginning of the 21st century large scale commercial use of biofuel is the most rapidly growing renewable energy source in the developed countries as well as developing countries. Several clean energy options are viable today and several others are likely to be so in the future, as technologies improve, costs are reduced, and the competitive landscape for biofuel technologies evolves.

The Intergovernmental Panel on Climate Change has considered a range of options for mitigating climate change and increased use of biomass for energy features in all of its scenarios. The biomass takes an increasing share of total energy over the next century, rising to 25±46% in 2100 in its five scenarios. In the biomass intensive energy scenario, with biomass providing for 46% of total energy in 2100, the target of stabilizing CO_2 in the atmosphere at present-day levels is approached. Annual CO_2 emissions fall from 6.2 Gt C in 1990 to 5.9 Gt C in 2025 and to 1.8 Gt C in 2100: this results in cumulative emissions of 448 Gt C between 1990 and 2100, compared to 1300 Gt C in their business-as-usual case (IPCC, 1996).

However, developing countries with tropical climates may have a comparative advantage in growing energy rich biomass and second generation technologies could enable expansion of the range of feedstock used from the traditional sugarcane, maize, and rapeseed to grasses and trees that can thrive in less fertile and more drought prone regions. Biodiesel production efforts are focused on using non-edible oil seeds from plants (*Jatropha curcas*, *Pongamia pinnata* and other tree borne oilseeds) and animal fats like fish oil. The focus is to encourage the use of wastelands and other unproductive land for the cultivation of these relatively hardy new biofuel crops so that biofuel feedstock crop cultivation does not compete with food crops for scarce agricultural land and water (Singh, 2009).

As ethanol yields improve or different feedstocks are introduced, ethanol production may become more economically feasible. Currently, research on improving ethanol yields from each unit of feedstock is underway using biotechnology. Also, as long as oil prices remain high, the economical use of other feedstocks such as cellulose, become viable. Environmental costs per unit of ethanol decline with higher biomass yield, lower fertilizer and fuel inputs into biomass production, and improvements in biomass to biofuel conversion efficiencies (Cassman and Liska, 2007). By-products such as straw or wood chips can be converted to ethanol. Fast growing species like switch grass can be grown on land not suitable for other cash crops and yield high levels of ethanol per unit area. The development of commercial cellulosic technology would allow agricultural residues to be used and increase ethanol yield per hectare.

The biogas plants may reduce the dependence on conventional sources of energy by the turn of the century, provided promotional efforts are continued. Although, cattle dung has been recognized as the chief raw material for bio-gas plants, other materials like night-soil, poultry litter and agricultural wastes are also used where they are socially acceptable.

Climate change, air quality and energy security will change the way energy is used and supplied over the next century. Supplying increasing amounts of clean and secure energy will be a challenge that will require a great deal of innovation and investment. There are plenty of biomass resource and technology options for biofuel productions that could lead to emissions reductions in the heat, transport and electricity sector, while improving energy security and air quality.

7. CONCLUSION

Biofuels are non-polluting, locally available, accessible and reliable fuels obtained from renewable sources. Biomass can act as a reservoir of carbon or as a direct substitute for fossil fuels with no net contribution to atmospheric CO_2 if produced and used sustainably. Fuel security and the reduction of air pollution are some of the fundamental gains of an expanded biofuels industry. When particularly favorable improvements in technology over the next decade are assumed, the costs of emissions from biofuel could be approximately equal to, but unlikely less than, those of conventional gasoline. Cellulosic ethanol holds the promise of yet greater environmental benefits, but economical ways of producing it must first be discovered. New biofuel feedstocks especially low input cultivation of non-food crops (e.g., *Jatropha*, hybrid poplar, new varieties of switchgrass, and better multispecies plant mixtures) and algal biodiesel production technology may also yield substantial improvements. Biofuel markets can serve as an opportunity to trigger additional investments that could lead to increased production of food as well as biofuel crops by small-scale farmers. Further research on the use of indigenous non-food crops should be encouraged.

Conversely, other ways of increasing biofuel production may increase air pollutant emissions unless accompanied by simultaneous improvements in abatement technology. Consideration should also be given to improved emissions controls and increases in fuel efficiency and fuel conservation that would reduce the need for increased fuel imports. Thus biofuels provide lots of environmental benefits including reduction of greenhouse gas emissions, improvement in air quality, reduction of fossil fuel use, increased national energy security, increased rural development and a sustainable fuel supply for the future and it also requires careful assessment on its impact of the environment especially in lowering greenhouse emissions.

REFERENCES

1. M. Akkerman, J. Janssen, Rocha, R. H. Wijffels, 2002Photobiological hydrogen production: photochemical efficiency and bioreactor design, International Journal of Hydrogen Energy, 2711951208
2. M. Al-Hasan, 2003Effect of ethanol-unleaded gasoline blends on engine performance and exhaust emission. Energy Conversion and Management, 4415471561
3. Y. Ali, M. A. Hanna, L. I. Lecviticus, 1995Emissions and power characteristics of diesel engines on methyl soyate and diesel fuel blends. Bioresource Technology, 52185195
4. Anonymous,2008Biofuels Research in the CGIAR: A Perspective from the Science Council. Position paper, 27Food and Aquiculture Organization of the United Nations FAO, Rome
5. Anonymous,2010aBiodiesel. Fuel fact sheets, 15.03.2010, Available fromhttp://www.biodiesel.org/pdf_files/fuelfactsheets/Lubricity.pdf
6. Anonymous,2010bGeneric biodiesel material safety data sheet, 15.03.2010, Available from http://www.biodiesel.org/pdf_files/fuelfactsheets/MSDS.pdf
7. A. Asankulova, A. D. Obozov, 2007Biogas in Kyrgyzstan. Applied Solar Energy 434262265
8. S. T. Bagley, L. D. Gratz, J. H. Johnson, J. F. Mc Donald, 1998Effects of oxidation catalytic converter and a biodiesel fuel on the chemical, mutagenic and particle size characteristics of emissions from a

diesel engine. Environmental Sciences and Technology, 3211831191

9. B. K. Barnwal, M. P. Sharma, 2005Prospects of biodiesel production from vegetable oils in India. Renewable and Sustainable Energy Review, 9363378

10. Battelle,1998Flammability limits for ethanol/diesel blends. Final Report prepared by Battelle, Columbus, OH, USA.

11. I. D. Bedoya, A. A. Arrieta, F. J. Cadavid, 2009Effects of mixing system and pilot fuel quality on diesel-biogas dual fuel engine performance. Bioresource Technology, 10066246629

12. Biogas Digest,2010Basics Information and Advisory Service on Appropriate Technology, Volume I. 15.11.2010, available from http://www.gtz.de/de/dokumente/en-biogas-1pdf

13. P. Borjesson, M. Berglund, 2006Environmental systems analysis of biogas systems-Part I: Fuel-cycle emissions. Biomass and Bioenergy, 305469

14. BSP,2000Biogas Support Program (BSP). Winrock International, Nepal, Biogas Support Program Nepal http://practicalaction.org/docs/energy/docs50/bp50-nepal-biogas.pdf

15. D. Budny, P. Sotero, 2007Brazil Institute Special Report: The Global Dynamics of Biofuels, Brazil Institute of the Woodrow Wilson Center, 03.05.2008, Available from http://www. Wilsoncenter .org/topics/pubs/Brazil_SR_e3.pdf

16. J. Bünger, J. Krahl, H. Franke, A. Munack, E. Hallier, 1998Mutagenic and cytotoxic effects of exhaust particulate matter of biodiesel compared to fossil diesel fuel. Mutation Research, 4151323

17. K. G. Cassman, A. Liska, 2007Food and fuel for all: realistic or foolish? Biofuels, Bioproducts and Biorefining, 11823

18. B. S. Chauhan, N. Kumar, S. S. Pal, D. J. Yong, 2011Experimental studies on fumigation of ethanol in a small capacity Diesel engine. Energy, 3610301038

19. D. Y. Z. Chhang, J. H. Van Gerpen, I. Lee, L. A. Johnson, E. G. Hammond, S. J. Marley, 1996Fuel properties and emissions of soybean oil esters as diesel fuel. J American oil and Chemical Society, 731549

20. Y. Chisti, 2008Biodiesel from microalgae beats bioethanol. Trends in Biotechnology, 26126131

21. D. R. Cohn, L. Bromberg, J. B. Heywood, 2005Direct Injection Ethanol Boosted Gasoline Engines: Biofuel Leveraging For Cost Effective Reduction of Oil Dependence and CO2 Emissions. April 20, 2005, available from http://ethanolboost.com/LFEE-200501 .pdfMassachusetts Institute of Technology, Cambridge, MA 02139

22. Coleman, Brooke2011A brief Summary of Air Quality and Impacts of Biofuels. Available from http://www .nebiofuels .org/pdfs/AQ_Summary.pdf

23. B. Delfort, I. Durand, G. Hillion, A. Jaecker-Voirol, X. Montagne, 2008Glycerin for new biodiesel formulation. Oil Gas Science and Technology Reveiw IFP, 634395404

24. J. Duailibi, 2008Ele é o falso vilão (in Portuguese). Veja Magazine, 03.05.2008, Available from http://veja.abril.com .br/ 300408/058s.html

25. T. G. Dunahay, E. E. Jarvis, S. S. Dais, Roessler, 1996P.G. Manipulation of microalgal lipid production using genetic engineering, Applied Biochemistry and Biotechnology, 57-58223231

26. T. D. Durbin, J. R. Collins, J. M. Norbeck, M. R. Smith, 2000Effects of biodiesel, biodiesel blends, and a synthetic diesel on emissions from light heavy-duty diesel vehicles. Environmental Science and Technology, 34349355

27. T. P. Durrett, C. Bennmg, J. Ohlrogge, 2008Plant triaglycerols as feedstocks for the production of biofuels. The Plant Journal 54593607

28. EIA,2010Energy Information and Administration, International Energy Statistics. 12.01.2010, Available from http://tonto.eia. doe.gov/cfapps/ipdbproject/IEDIndex3.cfm

29. N. El Bassam, 1998Energy Plant Species-Their Use and Impact on Environment, 321London, James & James, Science Publishers Ltd

30. M. A. Elsayed, R. Matthews, N. D. Mortimer, 2003Carbon and Energy Balances for a Range of Biofuels Options. Resources Research Unit, Sheffield Hallam University

31. Enagri,2010Global ethanol production to reach 85.9 bn litres. Enagri eMagazine, 47201750-6972Lincolnshire, UK

32. EPA,2009U.S. Environmental Protection Agency, EPA Proposes New Regulations for the National Renewable Fuel Standard

Program for 2010 and beyond, fact sheet, 3May 2009, Washington, DC

33. FAO,2005Bioenergy. Sustainable Development Department, FAO, Rome, Italy. 11.3.2006, Available from http://www.fao.org/sd/dim_en2/en2_050402_en.htm

34. FAO,2008FAO, The State of Food and Agriculture, Biofuels: Prospects, Risks and Opportunities, 978-9-25105-980-7Chapter 3, 3032

35. A. E. Farrell, R. J. Plevin, B. T. Turner, A. D. Jones, M. O'Hare, D. M. Kammen, 2006Science, 311506508

36. A. S. Fedorov, S. Kosourov, M. L. Ghirardi, M. Seibert, 2005Continuous H2 photoproduction by Chlamydomonas reinhardtii using a novel two-stage, sulfate-limited chemostat system, Applied Biochemistry and Biotechnology, 121-124403412

37. B. J. Finlayson-Pitts, Pitts Jr, J. N. , 2000Chemistry of the Upper and Lower Atmosphere: Theory, Experiments, and Applications. Academic Press, San Diego

38. L. Fulton, T. Howes, J. Hardy, 2004Biofuels for Transport: An International Perspective. International Energy Agency, Paris

39. J. S. Gaffney, N. A. Marley, 2009The impacts of combustion emissions on air quality and climate- From coal to biofuels and beyond, Atmospheric Environment, 432336

40. J. S. Gaffney, N. A. Marley, 2000Alternative fuels. In: Brimblecombe, P., Maynard, R. (Eds.), Air Pollution Reviews. The Urban Air Atmosphere and Its Effects, 1Chapter 6, 195246Imperial College Press, London, UK

41. J. S. Gaffney, R. Sapienza, T. Butcher, C. Krishna, W. Marlow, T. O'Hare, 1980Soot reduction in diesel engines: a chemical approach. Combustion Science and Technology, 248992

42. M. Gavrilescu, Y. Chisti, 2005Biotechnology- a sustainable alternative for chemical industry, Biotechnology Advances, 23471499

43. M. L. Ghirardi, J. P. Zhang, J. W. Lee, T. Flynn, M. Seibert, E. Greenbaum, 2000Microalgae: a green source of renewable H2. Trends in Biotechnology, 18506511

44. J. Goettemoeller, A. Goettemoeller, 2007Sustainable Ethanol: Biofuels, Biorefineries, Cellulosic Biomass, Flex-Fuel Vehicles, and

Sustainable Farming for Energy Independence, 42Prairie Oak Publishing, Maryville, Missouri

45. C. B. Granda, Zhu. Li, M. T. Holtzapple, 2007Sustainable liquid biofuels and their environmental Impact. Environmental Progress, 26233250

46. I. A. Guschina, J. L. Harwood, 2006Lipids and lipid metabolism in eukaryotic algae, Progress in Lipid Research, 45160186

47. D. O. Hall, J. I. House, 1993Reducing atmospheric CO, using biomass energy and photobiology. Energy Conversion Management, 34889896

48. HEI,1996The potential health effects of oxygenates added to gasoline: a review of the current literature. Health Effects Institute, Cambridge, MA. 01.01.1996, Available from http://www.healtheffects.org/Pubs/oxysum.htm

49. S. Hoekman, Kent, 2009Biofuels in the U.S.-Challenges and Opportunities. Renewable Energy, 341422

50. Q. Hu, M. Sommerfeld, E. Jarvis, M. Ghirardi, M. Posewitz, M. Seibert, A. Darzins, 2008Microalgal triacylglycerols as feedstocks for biofuel production: perspectives and advances. Plant Journal, 54621639

51. H. J. Huang, S. Ramaswamy, U. W. Tschirner, B. V. Ramarao, 2008A review of separation technologies in current and future biorefineries. Separation and Purification Technology, 62121

52. B. Hulsey, 2006Cleaning the Air with Ethanol, Reports Ethanol Today, Available from www.betterenvironmental solutions.com/reports/EthanolToday.pdf,5657

53. IEA-ETE,2007IEA-Energy Technology Essentials-Biofuel Production (ETE02), January. 2007, 14Available from http://www.iea.org/techno/essentials2.pdf

54. Indian express2008Petrol with 20% biofuel to be mandatory by 2017. http:/www.indianexpress.com/news/petrol-with-20biofuel-to be mandatory-by/333541/

55. IPCC,1996Climate Change 1995: Impacts, adaptations and mitigation of climate change: Scienti®c-technical analysis, Intergovernmental Panel on Climate Change Working Group II report, Cambridge University Press, Cambridge, 1996

56. M. Ishida, S. Yamamoto, H. Ueki, D. Sakaguchi, 2010Remarkable Improvement of NOx-PM trade-off in a Diesel Engine by Means of Bioethanol and EGR. Energy, 351245724581

57. M. Z. Jacobson, 2007Effects of Ethanol (E85) vs. Gasoline Vehicles on Cancer and Mortality in the United States. ACS Publications, 14.01.2008, available from http://pubs.acs.org/cgi-bin/sample.cgi/esthag/asap/html/es062085v.html

58. S. Jahangirian, A. Engeda, I. S. Wichman, 2009Thermal and Chemical Structure of Biogas Counter flow Diffusion Flames. Energy and Fuels, 2353125321

59. S. K. Jha, S. Fernando, E. Columbus, H. Willcutt, 2009A Comparative Study of Exhaust Emissions Using Diesel-Biodiesel-Ethanol Blends in New and Used Engines. Transactions of the American Society for Agricultural and Biological Engineering, 522375381

60. A. M. John, 2008Carbon-negative biofuels. Energy Policy, 36940945

61. I. K. Kapdan, F. Kargi, 2006Bio-hydrogen production from waste materials. Enzyme and Microbial Technology, 38569582

62. D. Keay, 2007Study warns of health risk from ethanol, San Francisco Chronicle, 18. 04. 2007, Available from http://sfgate.com/cgi-bin/article.cgi?file

63. S. Kim, B. E. Dale, 2004Global potential bioethanol production from wasted crops and crop residues. Biomass and Bioenergy 26361375

64. G. Knothe, K. R. Steidley, 2005Lubricity of components of biodiesel and petrodiesel: The origin of biodiesel lubricity. Energy and Fuels, 1911921200

65. G. Knothe, 2006Biodiesel and vegetable oil fuels: Then and now. Paper presented at the 97th American Oil Chemists Society Annual Meeting, St. Louis, MO

66. B. C. P. Koo, D. Y. C. Leung, 2000Emission testing on a biodiesel produced from animal fats. In: Proceedings of 3rd APCSEET, 2422469-81024-549-1Scientific Publishing, Singapore, Hong Kong

67. U. Krisztina, D. Scarpete, T. Panait, D. Marcel, 2010Thermo economical Performance Criteria in Using Biofuels for Internal Combustion Engines. Advances in Energy Planning, 8186

68. O. Kruse, J. Rupprecht, J. H. Mussgnug, G. C. Dismukes, B. Hankamer, 2005Photosynthesis: a blueprint for solar energy capture and

biohydrogen production technologies. Photochemical and Photobiological Science, 4957969

69. A. Kumar, S. Sharma, 2008An evaluation of multipurpose Oil seed crop for industrial uses Jatropha curcas L.): a review. Industrial Crops and Products 28110

70. Y. Lafay, B. Taupin, G. Martins, G. Cabot, B. Renou, A. Boukhalfa, 2007Experimental study of biogas combustion using a gas turbine configuration. Experimental Fluids, 432395

71. E. D. Larson, 2006A review of life-cycle analysis studies on liquid biofuel systems for the transport sector, Energy and Sustainable Development, 102109126

72. Y. Lechon, H. Cabal, C. de la Rua, N. Caldes, M. Santamarıa, R. Saez, 2009Energy and greenhouse gas emission savings of biofuels in Spain's transport fuel. The adoption of the EU policy on Biofuels. Biomass and Bioenergy, 33920932

73. J. S. Levine, 1991Global Biomass Burning: Atmospheric, Climatic, and Biospheric Implications. The MIT Press, 2530Cambridge, MA

74. L. R. Lynd, J. H. Cushaman, R. J. Nichols, C. E. Wyman, 1991Fuel ethanol from cellulosic biomass. Science, 2511318

75. F. , M. A. Hanna, 1999Biodiesel production: a review. Bioresource Technology, 70115

76. N. Marek, J. Evanoff, 2001The use of ethanol blended diesel fuel in unmodified, compression ignition engines: an interim case study. In: Proceedings of the Air and Waste Management Association, 94th Annual Conference and Exhibition, Orlando, FL

77. P. Meiring, A. C. Hansen, A. P. Vosloo, P. W. L. Lyne, 1983High concentration ethanol-diesel blends for compression-ignition engines. SAE Technical Paper 831360Society of Automotive Engineers, Warrendale, PA

78. F. B. Metting, 1996Biodiversity and application of microalgae, Journal of Industrial Microbiology, 17477489

79. P. Metzger, C. Largeau, 2005Botryococcus braunii: a rich source for hydrocarbons and related ether lipids, Applied Microbiology Biotechnology, 66486496

80. M. Mittelbach, C. Remschmidt, 2004Biodiesel, The comprehensive handbook, 2735Boersedruck Ges, M.B.H., Vienna, Austria

81. MNES2006Renewable energy for rural applications. Ministry of Non-conventional Energy Sources, Government of India

82. R. E. Morris, A. K. Pollack, G. E. Mansell, C. Lindhjem, Y. Jia, G. Wilson, 2003Impact of Biodiesel Fuels on Air Quality and Human Health Summary Report September 161999January 31, 2003. NREL/SR-540-33793

83. S. M. Mudge, G. Pereira, 1999Stimulating the biodegradation of crude oil with biodiesel preliminary results. Spill Science and Technology Bulletin, 5353355

84. Z. Nicholas, 2007Coproducts Energy Value is rising, Ethanol Producer Magazine, October 2007

85. NREL,2006From biomass to biofuels. National Renewable Energy Laboratory, NREL/BR-51039436

86. M. E. D. Oliveira, de , B. E. Vaughan, Rykiel Jr, E. J. , 2005Ethanol as Fuel: Energy, Carbon Dioxide Balances, and Ecological Footprint, BioScience, 557593602

87. D. C. Parashar, R. Gadi, T. K. Mandal, A. P. Mitra, 2005Carbonaceous aerosol emissions from India. Atmospheric Environment, 3978617871

88. H. Pathak, N. Jain, A. Bhatia, S. Mohanty, N. Gupta, 2009Global warming mitigation potential of biogas plants in India, Environmental Monitoring and Assessment, 157407418

89. J. G. Peter, J. G. David, N. S. John, 2003Wood-ethanol for climate change mitigation in Canada. Applied Biochemistry and Biotechnology, 1051-3231242

90. R. Petracek, 2011The Advantages of Using Biodiesel Blends. 26.02.2011, Available fromhttp://www.buzzle.com/articles/the-advantages-of-using-biodiesel-blends.html

91. J. Pitkanen, A. Aristidou, L. Salusjarvi, L. Ruohonen, M. Penttila, 2003Metabolic flux analysis of xylose metabolism in recombinant Saccharomyces cerevisiae using continuous culture. Metabolic Engineering, 51631

92. Planning Commission,2003Report of Committee on Development of Biofuel. (2003). Planning Commission, Government of India, New Delhi, India

93. B. Popa, 2010Emissions: Gasoline vs. Diesel vs. Bioethanol, 27.12.2010, Available fromhttp://www.autoevolution. com/

news/emissions-gasoline-vs-diesel-vs-bioethanol-3657.html, autoevolution.com

94. S. Prasad, A. Singh, H. C. Joshi, 2007aEthanol as an alternative fuel from agricultural, industrial and urban residues. Resources Conservation and Recycling, 50139

95. S. Prasad, A. Singh, N. Jain, H. C. Joshi, 2007bEthanol production from sweet sorghum syrup for utilization as automotive fuel in India. Energy and Fuels, 21424152420

96. P. S. Rao, K. V. Gopalakrishnan, 1991Vegetable oils and their methylesters as fuels for diesel engines. Indian Journal of Technology, 29292297

97. REN21.2008RE Policy Network for the 21st Century Global Status Report, Available from http://www.martinot.info/ RE2007 _Global_Status_Report.pdf

98. T. W. Ryan, 1999Characterization of vegetable oils for use as fuels in Diesel engines ASAE, 4/99, 1999

99. K. Sanderson, 2006A field in ferment, Business feature, Nature, 444673676

100. J. Sheehan, T. Dunahay, J. Benemann, P. Roessler, 1998A look back at the U.S. Department of Energy's Aquatic Species Program-biodiesel from algae, National Renewable Energy Laboratory, Golden, CO Report NREL/TP-58024190

101. John. J. Sheehan, 2009Biofuels and the conundrum of sustainability. Current Opinion in Biotechnology, 20318324

102. P. R. Shukla, 1997Energy Strategies and Greenhouse Gas Mitigation: Models and Policy Analysis for India. Allied Publishers, New Delhi

103. S. Singh, 2009India Biofuels. Annual Report IN9080USDA, Foreign Agricultural Service. Global Agri Info Network, Available from http://gain.fas.usda.gov/

104. Sinobioenergy,2011Approval for use in USA, Executive Summary 1. 26.02.11 available from http://www.sinobioenergy .com/enArticleView.aspx?id=99v

105. K. R. Smith, 1993Fuel combustion, air pollution exposure, and health: The situation in developing countries. Annual Review of Energy and Environment, 18529566

106. K. S. Smith, 1987Biofuels, Air Pollution, and Health, Plenum Publishers, New York.

107. C. Somerville, 2006The billion ton biofuels vision. Science, 3121277

108. H. K. Speidel, R. I. Lightner, I. Ahmed, 2000Biodegradability of new engineered fuels compared to econventional petroleum yuels and alternative fuels in current use. Applied Biochemistry and Biotechnology, 86879897

109. P. Spolaore, C. Joannis, E. Duran, A. Isambert, 2006Commercial applications of microalgae. Journal of Biosciences and Bioengineering, 1018796

110. N. Stauffe, 2006Pint-sized engine promises high efficiency. MIT Tech Talk 51618Available from http://web.mit.edu/newsoffice /2006/techtalk51-6.pdf

111. J. Stokes, T. H. Lake, R. J. Osborne, 2000A Gasoline Engine Concept for Improved Fuel Economy-The Lean Boost System. (Technical Paper) SAE paper 2001-01-2902

112. K. A. Subramanian, S. K. Singal, M. Saxena, S. Singhal, 2005Utilization of liquid biofuels in automotive diesel engines: An Indian Perspective. Biomass and Bioenergy, 296572

113. M. H. Tachinardi, 2008Por que a cana é melhor que o milho, (in Portuguese). Época Magazine, 73Available from http:// revistaepoca.globo.com/Revista/Epoca/0,,EMI5865-15273.html

114. D. Tilman, J. Hill, C. Lehman, 2006Carbon-negative biofuels from low-input high diversity grassland biomass. Science, 31415981600

115. Traffic & Public Transport Authority,2000Technology and biogas use in Sweden, November 2000, City of Gothenburg, Sweden

116. USDoE,2007US department of energy on greenhouse gases. 09.09.2007, Available http://www.eia.doe.gov/oiaf/1605/ ggccebro/chapter1.html

117. R. H. Vaivads, M. F. Bardon, V. K. Rao, V. Battista, 1995Flammability tests of alcohol/gasoline vapours. SAE Technical Paper 950401

118. P. Vasudevan, S. Sharma, A. Kumar, 2005Liquid fuel from biomass: an Overview, Journal of Scientific and Industrial Research, 64822831

119. G. Vellguth, 1983Performance of vegetable oils and their monoesters as fuels for diesel engines, Society of Automotive Engineers, SAE Technical paper series 831358Warrendale, PA, USA

120. M. Wang, C. Saricks, D. Santini, 2009Effects of Fuel Ethanol Use on Fuel-Cycle Energy and Greenhouse Gas Emissions, Argonne National Laboratory. 07. 07. 2009, Available from http://www.transportation.anl.gov/pdfs/TA/58.pdf

121. W. G. Wang, N. N. Clark, D. W. Lyons, R. M. Yang, M. Gautam, M. Bata, J. L. Loth, 1997Emissions comparisons from alternate fuel busses and diesel busses with a chassis dynamometer testing facility. Environment Science and Technology, 3131323137

122. Y. Zhang, M. A. Dube, D. D. Mc Lean, M. Kates, 2003Biodiesel production from waste cooking oil: 2. Economic assessment and sensitivity analysis. Bioresource Technology, 90229240

123. L. Zou, S. Atkinson, 2003Characterizing vehicle emissions from the burning of biodiesel made from vegetable oil. Environmental Technology, 241012531260

5

CHAPTER

BIOHYDROGEN PRODUCTION FROM LIGNOCELLULOSIC BIOMASS: TECHNOLOGY AND SUSTAINABILITY

Anoop Singh [1,*], Surajbhan Sevda [2], Ibrahim M. Abu Reesh [2], Karolien Vanbroekhoven [3], Dheeraj Rathore [4] and Deepak Pant [3,*]

[1] *Government of India, Ministry of Science and Technology, Department of Scientific and Industrial Research (DSIR), Technology Bhawan, New Mehrauli Road, New Delhi 110016, India*

[2] *Department of Chemical Engineering, Qatar University, Doha 2713, Qatar*

[3] *Separation and Conversion Technologies, Flemish Institute for Technological Research (VITO), Mol 2400, Belgium*

[4] *School of Environment and Sustainable Development, Central University of Gujarat, Gandhinagar 382030, Gujarat, India*

ABSTRACT

Among the various renewable energy sources, biohydrogen is gaining a lot of traction as it has very high efficiency of conversion to usable power with less pollutant generation. The various technologies available for the production of biohydrogen from lignocellulosic biomass such as direct biophotolysis, indirect biophotolysis, photo, and dark fermentations have some drawbacks (e.g., low yield and slower production rate, etc.), which limits their practical application. Among these, metabolic engineering is presently the most promising for the production of biohydrogen as it overcomes most of the limitations in other technologies. Microbial electrolysis is another recent technology that is progressing very rapidly. However, it is the dark fermentation approach, followed by photo fermentation, which seem closer to commercialization. Biohydrogen production from lignocellulosic biomass is particularly suitable for relatively small and decentralized systems and it can be considered as an important sustainable and renewable energy source. The comprehensive life cycle assessment (LCA) of biohydrogen production from lignocellulosic biomass and its comparison with other biofuels can be a tool for policy decisions. In this paper, we discuss the various possible approaches for producing biohydrogen from lignocellulosic biomass which is an globally available abundant resource. The main technological challenges are discussed in detail, followed by potential solutions.

Keywords: biohydrogen; biofuels; lignocellulosic biomass; technology; sustainability; life cycle assessment

1. INTRODUCTION

Recent years have seen a rapid surge in research activities focusing intensely on alternative fuels in order to reduce the dependency on fossil fuels, mainly by providing local energetic resources. This is mainly due to two reasons, the first being that new fuels are needed to supplement and ultimately replace depleting oil reserves and secondly, fuels capable of low or nil CO_2 emissions are urgently required to reduce the impact of global warming [1,2,3,4,5,6,7]. Hydrogen (H_2), which can be used in fuel cells mainly to operate machines, is a fascinating alternative, particularly because its combustion provides high amounts of energy and water is the only reaction product. Among all biofuels, H_2 has the highest gravimetric energy density at 141 MJ/kg. Despite this, its volumetric energy density, at only 12 MJ/m^3 (at normal temperature and pressure) is low. This is an important aspect particularly in reference to transportation fuel. It is considered to be one of the cleanest energy carriers if produced using energy generated from renewable sources. In summary, H_2 is interesting due to its potentially high efficiency of conversion to usable power, low generation of pollutants and high energy density [8]. Global H_2 production today amounts to around 700 billion Nm^3 and is based almost exclusively on fossil fuels [9]. However, for H_2 to be accepted as a sustainable substitute for fossil fuels, it has to be produced from renewable feedstock other than fossil fuels [10].

Hydrogen has been suggested as the ideal fuel of the future. It is considered as one of the cleanest energy carriers to be generated from renewable sources [11]. It has a high energy yield (122 kJ/g) which is 2.75 times greater than hydrocarbon fuels. It can be easily used in fuel cells for generation of electricity. Though not a primary energy source, it serves as a medium through which primary energy sources (such as H_2 produced from nuclear power and/or solar energy) can be stored, transported and utilized to fulfill our energy needs. The major problem facing H_2 as a fuel is its unavailability in Nature. H_2 can be produced safely, is environmentally friendly when combusted, and versatile i.e., has many potential energy uses, including powering non-polluting vehicles, heating homes and offices, and fueling aircraft. Current H_2 production technologies such as steam reforming of natural gas, thermal cracking or coal gasification are not environmentally friendly. Biological H_2 production is a promising alternative. There are two methods to produce H_2 from microorganisms. The first method uses photosynthetic microorganisms such as bacteria or algae (photofermentative processes) and the second method uses fermentative organisms (dark fermentation processes). Fermentative H_2 production has the advantage of producing H_2 under mild conditions with the additional benefit of allowing residual biomass valorization. The dark fermentation process is more attractive as it has the potential to use wastewater and organic wastes and has higher production rates compared to photofermentative processes. So far, few studies have used real wastewater for the production of H_2 due to inhibition by both substrate and/or product in the fermentation process [12]. Studies on bioH_2 production have been focused on photodecomposition of organic compounds by photosynthetic bacteria, dark fermentation from organic compounds with anaerobes and biophotolysis of water using algae and cyanobacteria [13,14,15,16,17].

Lignocellulosic biomass is the most abundant in Nature and it is present in hardwood, softwood, grasses, and agricultural residues. The global annual yields of lignocellulosic biomass residues were estimated to exceed 220 billion tons, equivalent to about 60–80 billion tons of crude oil [12]. Lignocellulosic feedstocks consist mainly of glucose and xylose and thus microbial strains that can effectively degrade glucose and xylose are important for development of renewable H_2 production processes [18]. Direct conversion of lignocellulosic biomass to H_2 needs pretreatment to hydrolyze the incorporated heterogeneous and crystalline structure [19,20]. The lignocellulosic biomass hence presents an attractive, low-cost feed stock for H_2 production.

Aim of the Paper

In recent past, several reviews have appeared which have discussed the prospects and challenges of biomass-based H_2 [21,22]. Earlier, Kraemer and Bagley gave a thorough description of the yield improvement approaches in fermentative H_2 production [23]. Wang and Wan summarized the main factors influencing fermentative H_2 production [24]. A special issue of the journal International Journal of Hydrogen Energy, recently dealt with "Biohydrogen: From Basic Concepts to Technology" [25]. Biohydrogen can be generated by adopting different technologies and different technologies can perform differently. Thus, the aim of this paper is to discuss specifically the technological aspect of biohydrogen production from lignocellulosic biomass and its sustainability on the basis of a life cycle assessment (LCA).

2. FEED STOCK FOR BIOHYDROGEN PRODUCTION

Glucose is the ideal substrate, but it is too costly at present. Many agricultural residues and food wastes are rich in carbohydrates that could serve as feedstock. Lignocellulosic biomass is another sustainable feedstock for H_2 production [26]. The criteria for an ideal feedstock for sustainable H_2 production, which include high carbohydrate content, minimum pre-treatment requirement, sustainable resources, low cost and sufficient concentration of carbohydrate for fermentative conversion, have been suggested by Bartacek [27]. The substrates usable for fermentative H_2 production were further divided into four main groups, namely, pure substrates such as glucose; energy crops such as Miscanthus, solid wastes like food waste and industrial wastewaters such as wastewater from the pulp and paper industry.

A variety of substrates has been used as feedstock for H_2 production. For example, the fermentation of household wastes under different temperature conditions has been well studied [28,29,30]. An increase in H_2 yields as temperature increased to thermophilic regimes was reported.

Wastewaters and residual biomass with high carbohydrate content have also been demonstrated to be a suitable candidate for dark fermentation. This includes molasses [31,32] and cheese whey [33,34], which have been evaluated under continuous stirred tank reactor (CSTR) and immobilized system configurations. Besides, H_2 production from soluble and particulate starch and cellulose [35,36], xylose [37], sugar beet [38], wastewater from a sugar beet refinery [39] and the bottom layer from a beer manufacturing plant [40] has also been demonstrated.

3. TECHNOLOGY

3.1. Biohydrogen Production Systems

The conventional methods for producing H_2 gas include steam reforming of methane and hydrocarbons, non-catalytic partial oxidation of fossil fuels and autothermal reforming. However, most of these methods are energy intensive processes requiring high temperatures (>850 °C). A general scheme of H_2 production from renewable sources is shown in Figure 1. Biological methods of H_2 production are preferable to chemical methods because of the possibility to use sunlight, CO_2 and organic wastes as substrates for environmentally benign conversions, under moderate conditions.

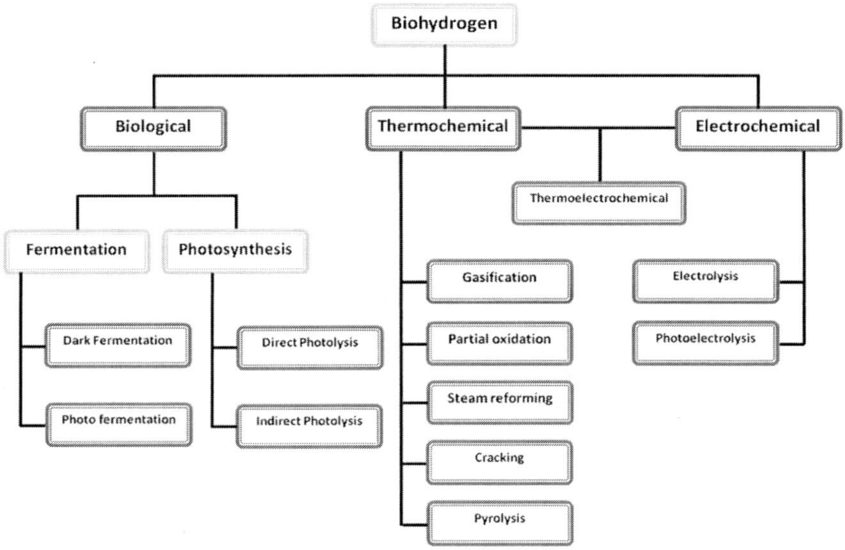

Figure 1. The main alternative methods of H_2 production from energy sources.

The biological production of H_2 involves light-dependent methods: direct and indirect biophotolysis, and photo fermentation. The other routes are light-independent methods, including the dark fermentation process and water-gas shift reaction of photoheterotrophic bacteria. This biologically produced H_2, generally referred to as "biohydrogen", is characterized by low H_2 yields which present a challenge for commercial applications.

3.1.1. Dark/Anaerobic Fermentation

Dark fermentation is one of the most common processes for bio-H_2 production. Although only 15%–20% of the theoretical H_2 potential of carbohydrates can be harvested, dark fermentation is considered as a promising process in a two phase anaerobic treatment system [23]. Also, since the CO_2 produced in dark fermentation has already been fixed by the waste treated originally from the atmosphere, the emissions associated with global climate change are virtually zero [41]. Microbial species analysis of hydrogen-producing cultures (using anaerobic sludge as inoculum) shows the presence of Clostridium cellulosi, Clostridium acetobutylicum, Clostridium tyrobutyricum, Enterobacteriaceae and Streptococcus bovis [42] Fermentative H_2 production usually proceeds from the anaerobic glycolytic breakdown of sugars. The theoretical complete oxidation of 1 mole of hexose to CO_2 can produce 12 moles of H_2. Nevertheless, the theoretical yield of H_2 via acetic acid fermentation cannot be higher than 4 moles. Actual H_2 yields are quite lower, typically ranging from 1.0 to 2.5 moles per mole of hexose consumed. Recently, Varanasi et al. reported production of 2.95 mol H_2/mol hexose equivalent by thermophillic dark fermentation using cellubiose as substrate [43]. If butyric acid is produced as the major fermentation product instead of acetic acid, only 2 moles of H_2 can be produced [44]. H_2 yield is even lower when more reduced organic compounds such as lactic acid, propionic acid and ethanol are produced, because these metabolites represent end products of metabolic pathways that bypass the major H_2-producing reaction [45]. Recently, it was concluded that to maximize net energy gain via dark fermentation, appropriate cultures capable of high-H_2 yield have to be employed and the process has to be operated at near-ambient temperatures with the lowest feedstock concentration as possible [10]. In an experiment with Thermotoga neapolitana sparged with N_2 and supplemented with 40 mM sodium bicarbonate a 2.8 and 2.7 mol/mol glucose yield of hydrogen with a lactic acid/acetic acid ratio of 0.26 was obtained, challenging the currently accepted dark fermentation model that predicts reduction of this gas when glucose is converted into organic products different from acetate [46]. Pradhan et al. reviewed the hydrogen production efficiency of a similar bacterium (Thermotoga neapolitana) with different feedstocks and found 1.9–3.5 mol H_2/mol hexose yields achievable with a range of feedstocks and variable substrate loads [47]. Byproducts of the reactions are acetic acid, lactic acid and ethanol.

3.1.2. Photo Fermentation

Photo fermentation is carried out by purple non-sulfur (PNS) photosynthetic bacteria which can grow as photoheterotrophs, photoautotrophs or chemoheterotrophs [48]. These bacteria produce H_2 under photoheterotrophic conditions (light, anaerobiosis, organic electron donor) [49]. The advantages of this process over photolysis of water using green algae and cyanobacteria, are that oxygen does not inhibit the process and that these bacteria can be used in a wide variety of conditions (i.e., batch processes, continuous cultures, and immobilized systems) [50]. The hydrogenase and nitrogenase enzymes produced in photosynthesis by green algae and photosynthetic bacteria, respectively, play a crucial role in biohydrogen production. The main PNS bacteria that participate in H_2 production are Rhodospirillum rubrum, Rhodopseudomonas palustris, Rhodobacter sphaeroides O.U 001, Rhodobacter sphaeroides RV, Rhodobacter sulfidophilus and Rhodobacter capsulatus. Kapdan et al. used three different pure strains of Rhodobacter sphaeroides (RV, NRLL and DSZM) in batch experiments to select the most suitable strain [51]. R. sphaeroides RV resulted in the highest cumulative hydrogen gas formation (178 mL), hydrogen yield (1.23 $mol{\cdot}H_2{\cdot}mol^{-1}$ glucose) and specific hydrogen production rate (46 $mL{\cdot}H_2{\cdot}g^{-1}$ $biomass{\cdot}h^{-1}$) at 5 $g{\cdot}L^{-1}$ initial total sugar concentration among the other pure cultures. Using Rhodobacter capsulatus JP91, Keskin and Hallenbeck compare the photofermentative biohydrogen yield of different feedstocks in a batch culture experiment [52]. Overall yield of biohydrogen was 10.5, 8 and 14.9 mol H_2/mol sucrose using beet molasses, black strap molasses and sucrose respectively. Optimization of process parameters such as availability of solar light, bioreactor configuration and proper C/N ratio in substrate (synthetic and derived from waste products) still needs to be studied at higher scale.

3.1.3. Combined Biotechnologies

Combination of two or more of the abovementioned techniques have also been studied for improved H_2 yields. Theoretically 12 moles of H_2 per mole of glucose can be generated by combining dark fermentation with photo fermentation (using PNS bacteria) [48]. For instance, Nath et al. studied combined dark and photo fermentation using glucose as substrate [53]. The effluent from the dark process (containing unconverted metabolites, mainly acetic acid) underwent photo fermentation by Rhodobacter sphaeroides in a column photo-bioreactor demonstrating the feasibility of this combination to achieve higher yields of H_2 by complete utilization of the chemical energy stored in the substrate. A sequential process using glucose as substrate and an immobilized system for the photo fermentation step evaluating key factors such as diluted ratio of dark fermentation effluent, ratio of dark and photo fermentation bacteria, light intensity, and light/dark cycle has also been studied [54]. During the combined process, maximum total H_2 yield was 5.374 moles of H_2/moles of glucose. However, the sterilization step applied to the dark fermentation effluent may pose a constraint to a scale-up the process. The combined system can also be run in continuous mode and achieve more combined H_2 yield [49]. These further combinations will reduce the overall cost of H_2 production but more field studies are required to obtain an economical H_2 production process.

3.1.4. Bioelectrochemical Production

Bioelectrochemical production of H_2 is the latest technology using systems called microbial electrolysis cells (MEC). This is an emerging field where the oxidation of organic material is carried out by the bacteria present at the anode and results in formation of protons, CO_2 and electrons (Figure 2). Protons migrate through a proton exchange membrane (PEM) to the cathode and the electrons are transported through the external circuit to the cathode [55]. By applying an external voltage of approximately 0.5–0.9 V, these electrons combine at the cathode with protons producing H_2 gas (Table 1). The advantage here is the low energy consumption (0.3–0.9 V) necessary for microbial electrohydrogenesis to produce H_2 in comparison to the theoretical minimum voltage of 1.23 V required for water electrolysis [56]. An overall scheme of H_2 production from lignocellulosic biomass is shown in Figure 3. Figure 3 summarized the pathways for biohydrogen production by using lignocellulosic biomass, depending on the nature of feed (solid or liquid), pre-treatment methods were used and then followed by the dark fermentation.

Table 2 shows the various substrates used for H_2 production in different MEC volumes. The rate of H_2 production differs with substrate due to their degradation pathways. To date MECs have shown H_2 production from initial volumes ranging from 5 mL to 1000 L (pilot plant) reactors, which shows that MECs can be a better solution for producing H_2 in the cathodic chamber while treating wastewater in the anodic chamber [58]. Still, there are many issues to be addressed for the long term real time application such as electrode and membrane stability for longer duration and reactor configuration design for higher volumes. Before scale up, mathematical models are required, which need to be validated first for the present lab scale studies and then on the basis of data obtained, higher volume MECs may be designed and validated.

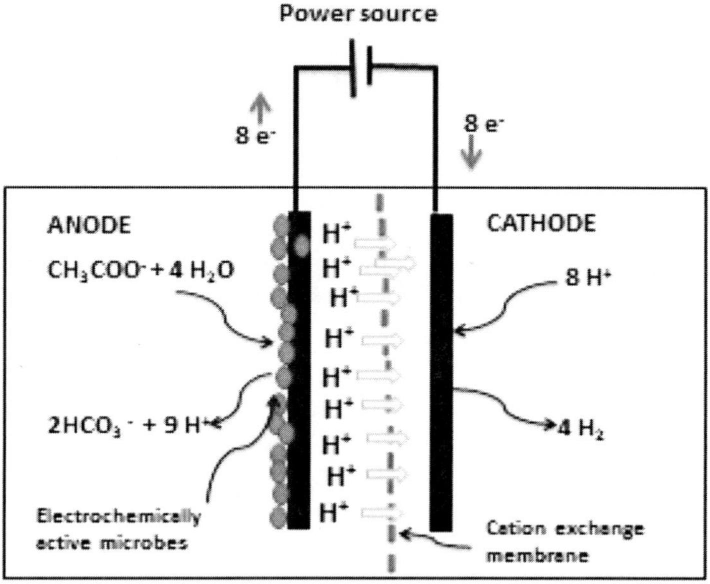

Figure 2. Schematic of hydrogen production in MEC (Adapted from [56]).

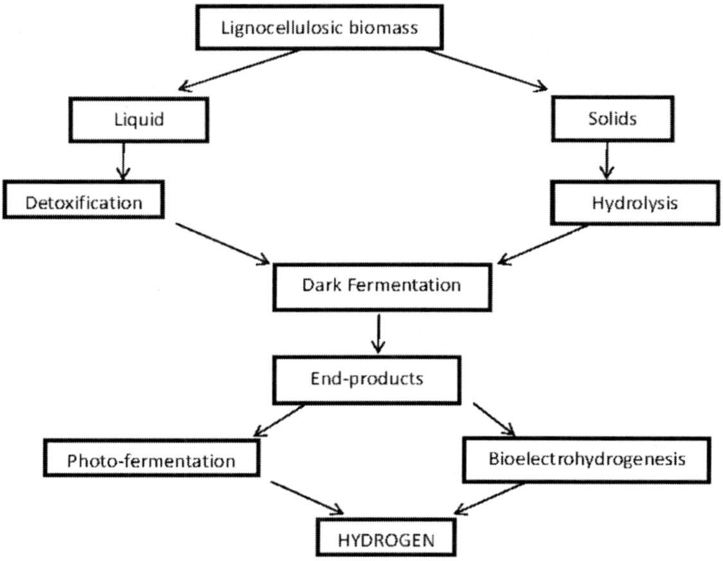

Figure 3. General scheme for biohydrogen production from lignocellulosic biomass (adapted from [12]).

Table 1. A comparison between the three major routes for biological hydrogen production (adapted from [57]).

Production Routes	Main Reaction	H_2 Production Rates (mmol/h· L)	Remark
Direct photolysis	$2H_2O$ + "light energy" → $2H_2 + O_2$	0.07	Similar to the processes found in plants and algal photosynthesis.
Photo fermentation	$C_6H_{12}O_6 + 6H_2O$ + "light energy" → $12H_2 + 6CO_2$, $\Delta G_0 = +3.2$ kJ	145–160	Bacteria evolve molecular H_2 catalyzed by nitrogenase under N-deficient conditions using light energy and reduced compounds (organic acids).
Dark fermentation	Pyruvate + CoA → acetyl-CoA + formate OR Pyruvate + CoA + 2Fd(ox) → Acetyl-CoA + CO_2 + 2Fd (red)	77	H_2 is produced by anaerobic bacteria, grown in the dark on carbohydrate rich substrate.

Table 2. Various substrates used in MEC for hydrogen production (adapted from [59]).

Substrate	Concentration (g/L)	Applied Voltage (V)	MEC Volume (mL)	Hydrogen Production Rate (m^3 H_2/m^3/day)	Reference
A de-oiled refinery wastewater	04–1	0.7	5	79% (Hydrogen production based on COD removal)	[60]
Sodium Acetate	1	0.6	18	2.0	[61]
Glucose	2	0.6	26	0.25 ± 0.03	[62]
Glucose	2	0.8	26	0.37 ± 0.04	[62]
Fermentation effluent	1	0.6	26	1.41	[63]
Sodium Acetate	1	0.6	28	1.99 ± 0.02	[64]
Sodium Acetate	1	0.8	28	3.12 ± 0.002	[64]
Sodium Acetate	1	0.5	28	1.7	[65]
Glucose	1	0.5	28	0.83 ± 0.18	[66]
Glucose	1	0.9	28	1.87 ± 0.30	[66]
Potato wastewater	1.9–2.5 (COD)	0.9	28	0.74	[67]
Swine wastewater	2 (COD)	0.5	28	0.9–1.0	[68]
Sodium Acetate	1	0.6	48	0.76	[65]
Sodium Acetate	1	0.7	76	-	[69]
Sodium Acetate	1	0.8	240	0.0231 ± 0.003	[70]
Sodium Acetate	1	1	400	1.58	[71]
Sodium Acetate	2	0.6	500	0.53	[72]
Winery wastewater	8	0.9	1000 Lt	0.19 ± 0.04	[58]
Sodium Acetate	1	0.5	6600	0.02	[56]

3.2. Microbiology of Biohydrogen Production

Perera et al. evaluated three main routes for biological H_2 production [10]. These are (1) direct photolysis, in which cyanobacteria decomposes water to generate hydrogen and oxygen in presence of light; (2) photo fermentation, where anoxygenic photoheterotrophic bacteria utilizes organic feedstock to produce H_2 in presence of light and (3) dark fermentation, in which anaerobic heterotrophic bacteria utilizes organic feedstock without any light to produce H_2. A comparison of these three main routes is shown in Table 1.

The microbiology and biochemistry of dark fermentative H_2 production was discussed in detail by Hawkes et al. [42]. H_2 production in Clostridia is due to the presence of hydrogenase enzymes. These transfer electrons from reduced ferredoxin or NADH to protons to regenerate the oxidized forms (Fd_{ox} and NAD+) required so that glycolysis and oxidative decarboxylation of pyruvate can proceed to generate ATP.

Pure microbial cultures have mainly been used in lab-scale reactors for studying the effect of environmental and operational parameters on fermentation profiles and carbon metabolism. One of the successful tests using pure culture in a pilot-scale bioreactor using a non-sterilized feedstock employed Caldicellulosiruptor saccharolyticus [73]. However, most studies on H_2 production on biowaste have been performed using mixed cultures under mesophilic conditions [74,75]. Only a few studies have focused on mixed thermophilic consortia [76,77]. It has been demonstrated that the extreme thermophile C. saccharolyticus can produce H_2 from mono- and disaccharides [78]. Hexose is the predominant component in the cellulose hydrolysates. A highest H_2 yield of approximately 83% of the theoretical value (4 $mol \cdot mol^{-1}$ hexose) has been reported using thermophilic anaerobic bacteria [78].

3.3. Limiting Factors in Biohydrogen Production Systems

The most challenging barrier of fermentative H_2 production is its low H_2 molar yield [26]. Thauer et al. predicted that 4 moles of H_2 per mole of glucose is the biological maximum in Clostridial microbes if acetate is the only waste by-product [79]. In practice, even that figure is rarely achieved. A number of factors adversely affect and inhibit H_2 fermentation [44]. H_2 itself, when it reaches high concentrations not only makes its production thermodynamically unfavourable but also acts as an inhibitory agent as do other metabolic products, such as acetic acid and propionic acid [17,80]. Partial pressure of H_2 is one of the most critical parameters in fermentative production of H_2 as high H_2 partial pressures make H_2 production thermodynamically unfavourable. Removal of produced H_2 from the liquid phase lowers the H_2 partial pressure which in turn increases H_2 yield [81]. Moreover, the H_2 remaining in the system might be consumed by some bacteria [82]. Removal of dissolved H_2 and reduction of H_2 partial pressure can be achieved by nitrogen flushing, adsorption of H_2 by metals and H_2 stripping by boiling or by introduction of steam [83,84,85]. Low H_2 partial pressure also needs to be maintained because hydrogenases (such as NiFe-hydrogenase) may re-oxidize the produced hydrogen into protons and electrons [86]. Gas sparging has proved to be an efficient method to maintain maximum hydrogen production even though it leads to biogas dilution and higher cost for hydrogen recovery [87]. Depending on the nature of the flushing gas, the flow rate and the reactor configuration, volumetric production of biogas up to 120% has been achieved [85,88]. Non-sparging techniques such as headspace modification under vacuum, high pressure or gas adsorption (reviewed in [87]), hydrogen-separating membranes [83] and using mechanical stirring [89,90], have also showed significant improvements in hydrogen yield. Argon has been often used to flush both oxygen and nitrogen and to keep a low H_2 partial pressure in the reactors, but it increases production costs and hinders H_2 purification [91]. Some researchers have reported reduced pressure and CO_2 for flushing the headspace and maintaining low H_2 partial pressure in dark fermentation [92,93], but the information on photofermentation is deficite. Montiel-Corona et al. suggested that flushing with Ar could be replaced with reduced pressure, which can be less expensive and practical for hydrogen recuperation [91]. Coupling the dark and photo fermentation showed an increased total hydrogen yield. One of the major drawbacks in coupling the dark and photo fermentation processes is the need of keeping apart the H_2-producing microflora and the presence of NH_4^+, which may be naturally present in wastewater and may also be generated in the dark fermentation process when hydraulic retention time (HRT) is high enough to achieve protein degradation, especially when particulate substrates (as in the case of food wastes) are being considered, since HRT may be as high as 5 days [94].

In case of bioelectrochemical production of H_2 in MEC, the main challenge is avoiding methane formation via methanogenesis [24], though researchers are now shifting more towards methane formation rather than H_2 in these systems [95,96]. Another issue limiting the large-scale application of this technology is the use of precious metal catalyst such as platinum which is usually used on the cathode [97]. Though there have been efforts to use low cost materials such as stainless steel [98] and Ni-based electrodes [61], the results are much lesser from the targets.

3.4. Role of Metabolic Engineering

The application of genomic and molecular tools has made it possible to steer the metabolic pathways towards maximal H_2 production and avoid waste and by-product accumulation. This is especially true when genetic engineering is conducted on cellulolytic microbes [26]. The main principles of genetic engineering include: (1) overexpression of cellulases, hemicellulases and lignases to maximize substrate availability, (2) elimination of H_2-consuming hydrogenases and (3) overexpression of H_2-producing hydrogenases [53]. Metabolic engineering modifications have been used to increase H_2 production in fermentative systems [99]. These include over-expression of H_2-evolving enzymes [100], the knockout of metabolic pathways that compete for reducing equivalents [81] and the introduction/over-expression of genes (cellulases, hemicellulases and lignases) to enhance carbohydrate availability to the cell [101]. Inactivation of the gene lactate dehydrogenase (ldhA) in E. coli by introducing mutations could lead to a modest increase (20%–45%) in net hydrogen production (reviewed in [102]).

Ryu et al. combined several known approaches to construct a superior hydrogen-producing strain of the purple nonsulfur photosynthetic bacterium Rhodobacter sphaeroides HPCA* (mutant expressing NifA L62Q) [103]. In this strain maximum hydrogen levels are reached almost twice as fast as in wild type cells and final hydrogen levels are ~39% higher than in the wild type as well. As increased number of genomes for H_2 producing microorganisms are sequenced and compared and as more specific enzymes are functionally characterized, the distinctive metabolic strategies used and enzymological contexts through which H_2 evolution is controlled in different organisms will become clearer. This will allow researchers to construct more effective strategies to modulate competing pathways, and help in the designing molecular engineering strategies leading to enhanced H_2 evolution.

4. KINETIC MODELS FOR HYDROGEN PRODUCTION BY FERMENTATION

Different factors such as substrate and inhibitor concentrations, temperature, pH and reactor type affect H_2 production by fermentation. Modeling of the H_2 production is very important to improve, analyze and predict H_2 production during fermentation. Mathematical models include the kinetic of cell growth and product(s) formation, substrate utilization and inhibition. In addition some models are developed to describe the effect of pH, temperature and dilution rate on H_2 production. The obtained model kinetic constants can be used in the design, operation and optimization of the fermentative H_2 production process. Different kinetic models have been proposed to describe growth of H_2 producing bacteria, substrate degradation and H_2 production. H_2 production is reported as growth associated product.

Monod (or Michaelis–Menten equation) (Equation (1)) which is an unstructured, non-segregated model of microbial growth, fits a wide range of data. The kinetic constants of this equation, K_S and μ_{max}, can be obtained by linear regression. Wang and Wan reported on previous studies using a Monod model to describe H_2 production with time in bio-H_2 fermentation [104]:

$$\mu = \frac{1}{X}\frac{dX}{dt} = \mu_m \frac{S}{K_s + S} \tag{1}$$

where μ is the specific growth rate, X is the biomass concentration, S is the substrate concentration, K_s is the saturation constant, μ_m is the maximum specific growth rate.

Recently, the logistic model (Equation (2)) became the most popular in describing cell growth. This equation has a sigmoidal shape that includes the lag phase, exponential and stationary phase of the batch growth:

$$\mu = \frac{1}{X}\frac{dX}{dt} = \mu_m \left(1 - \frac{X}{X_m}\right) \tag{2}$$

where X_m is the maximum biomass concentration.

At high substrate concentration, the cell growth is inhibited and production of H_2 is reduced. Different substrate inhibition models have been proposed. The Haldane-Andrew model (Equation (3)) is widely used to describe the substrate dependence of the specific growth rate of H_2 fermentations. Wang and Wan have reported that previous studies used an Andrews model to describe H_2 production with time [104]. Other substrate inhibition models are used in the literature such as modified Han-Levenspiel model (Equation (4)):

$$\mu = \frac{1}{X}\frac{dX}{dt} = \mu_m \frac{S}{K_s + S + \frac{S^2}{K_i}} \tag{3}$$

where K_i is the inhibition constant.

The presence of other inhibitors such as salts and the product cause reduction of H_2 production. Some models have been proposed to describe the effect of inhibitors such as the modified Han-Levenspiel model (Equation (4)):

$$\mu = \frac{1}{X}\frac{dX}{dt} = \mu_m \left(1 - \frac{C}{C_m}\right) \quad (4)$$

where C is the inhibitor concentration, C_m is the maximum inhibitor concentration or the concentration of inhibitor above which there is no biomass growth

The modified Gompertz model (Equation (5)) is widely used to describe the progress of cumulative H_2 production in batch fermentations [104]:

$$H_t = H_{max} \exp\left\{-\exp\left[\frac{R_{max} \times e}{H_{max}}(\lambda - t) + 1\right]\right\} \quad (5)$$

where H_t is the cumulative volume of H_2 produced at any time (mL), H_{max} is the gas production potential (mL), R_{max} is the maximum gas production rate (mL/h), λ is the lag time (h). t is the incubation time (h)

The Luedeking-Piret model (Equation (6)) has been widely used to describe the relation between cell growth rate and H_2 production:

$$\frac{dP}{dt} = Y_{P/X}\frac{dX}{dt} + \beta X \quad (6)$$

where P is the product, $Y_{P/X}$ is the growth associated yield coefficient; β is the non-growth associated product yield coefficient.

Wang and Wan reported that previous studies used the Luedeking–Piret model to relate cell growth rate and H_2 production rate [104]. The effect of temperature on the fermentative H_2 production has been widely described using the Arrhenius model, while the effect of pH on the substrate consumption rate is described by an Andrew model using the concentration of H^+ as the limiting substrate concentration. According to this model, the rate of substrate consumption passes through maximum with increasing H^+ concentration.

5. SUSTAINABILITY AND LIFE CYCLE ASSESSMENT

The concept of sustainable development is an attempt to combine growing concerns about a range of environmental issues with socio-economic issues and implies smooth transition to more effective technologies from a point view of an environmental impact and energy efficiency [105,106]. H_2 can be considered one of the pillars of a future sustainable energy system [107]. H_2

Table 3. Comparison of different biohydrogen production systems.

Reactor	Feed Stock	Maximum H_2 Yield	Reference
		Fermentation	
		Dark fermentation	
CSTR	Starch	0.52 L/h/L and 13.2 mmol H_2/g total sugar	[113]
Batch	Glycerol	0.41 mol H_2/mol glycerol	[114]
FBR	Sucrose	4.26 mol H_2/mol sucrose	[115]
Batch	Food waste	593 mL H_2/g carbohydrate	[116]
Fed-batch	Swine manure	18.7×10^{-3} g H_2 per g TVS	[117]
Batch	Sucrose	4.3 mol H_2/mol sucrose	[118]
Batch	Fructose, sorbitol, glucose	1.27, 1.46 and 1.51 mol H_2/substrate	[119]
Fed-batch	Starch, glucose	465 mL H_2/g starch, 3.1 mol H_2/mol glucose	[120]
Batch	Food waste	39.14 mL H_2/g food waste (219.91 mL H_2/VS_{added})	[121]
Batch	Crude Glycerol	64.24 mmol H_2/L and 5.74 mmol H_2/g COD consumed	[122]
Batch	Distillery wastewaters	1 L H_2/L medium	[123]
Batch	Cheese whey	94.2 L H_2/kg_{VS}	[124]
Batch	Water hyacinth (leaves and stems)	76.7 mL H_2/g TVS was obtained at 20 g/L of water hyacinth	[125]
Batch	waste ground wheat solution	SHPR = 25.7 mL H_2/g cells/h	[126]
		Photo fermentation	
CSTR	Sucrose	5.81 mol H_2/mol hexose	[127]
Fed-batch operation	Wheat starch	201 mL H_2 g/L starch	[128]
Batch	Molasses	0.50 mmol H_2/L_c h	[129]
Batch	Beet molasses	10.5 mol H_2/mol sucrose	[52]
Batch	Black strap	8 mol H_2/mol sucrose	[52]
Batch	Sucrose	14 mol H_2/mol sucrose	[52]
Batch	Ground wheat starch	46 mL H_2/g biomass/h, 1.23 mol H_2/mol glucose	[51]
Batch	lignocellulose-derived organic acids	7 mL H_2/mL of the fermentation effluent	[130]
		Photosynthesis	
		Direct Photolysis	
Batch	Lactate	0.07 mmol H_2 (l × h) or 54 mL/h· g dry weight	[131]
		Indirect Photolysis	
Batch	arabinose and xylose	14.55 mmol/g (arabinose); 13.73 mmol/g (xylose)	[132]
		Thermochemical	
		Gasification	
Continuous supercritical water gasification	glucose	10.5–11.2 mol/mol glucose	[133]
		Partial Oxidation	
Batch	municipal sludge	Not reported the amount	[134]
		Steam reforming	
molten carbonate fuel cell (MCFC) system	ethanol	5 mol H_2/mol fed ethanol	[135]
		Cracking	
fixed-bed quartz micro reactor	Methane	500 μmoles/min	[136]
		Pyrolysis	
stainless steel tank reactor	Biomass (redwood sawdust; cole stalk and rice husk) feed	65.39 g/Kg biomass for redwood sawdust; 40.0 g/Kg biomass for cole stalk and rice husk	[137]
		Thermoelectrochemical	
membrane electrode assembly	sulfur dioxide	0.4 A/cm^2 at 0.835 V (H_2 production rate did not reported)	[138]
membrane electrode assembly	anhydrous hydrogen bromide	2.0 A/cm^2 at 1.91 V (H_2 production rate did not reported)	[138]

Table 3. *Cont.*

Reactor	Feed Stock	Maximum H_2 Yield	Reference
	Electrochemical		
	Electrolysis		
The $BiOx-TiO_2$ electrode and stainless steel (SS, Hastelloy C-22) were used as an anode and a cathode in the electrochemical system, respectively	arsenite (As(III))	9.4 μmoles/min	[139]
	Photoelectrolysis		
The TiO_2(ns) was prepared in the form of a sol-gel	photoelectrode system TiO_2(ns)–VO_2	6 $L \cdot h^{-1} \cdot m^{-2}$ for the TiO_2(ns); 13.0 $L \cdot h^{-1} \cdot m^{-2}$ for the TiO_2(ns)–VO_2 photoelectrode	[140]

production could be a possible avenue for the large-scale sustainable generation of H_2 needed to fuel a future H_2 economy [106]. Despite its many obvious advantages, there remains a problem with storage and transportation. Pressurized H_2 gas occupies a great deal of volume compared with other fuels. For example, gasoline that with equal energy content, needs about 30 times less volume at 100 bar gas pressure. Due to its high explosivity there are also obvious safety concerns with the use of pressurized or liquefied H_2 in vehicles as well as additional energy use for pressurizing or liquefaction. Furthermore, the overall energy balance of using H_2 as vehicle fuel does indeed seem to be less beneficial than gasoline, but being the only non-carbon fuel it may still make sense to produce H_2 from waste streams if some of the obstacles can be solved and it can be used effectively for energy production to feed into grid or to use in stationary requirements, e.g., industries, etc.

Though this paper is focused on bio-H_2 production from lignocellulosic biomass, it is important to compare it to other production methods using various substrates. Such a comparison has been made in Table 3 by presenting the various H_2 production systems, which show different H_2 yields from different feedstocks by adopting different production systems. Therefore, life cycle assessment (LCA) could be a tool to scrutinize the best H_2 production system for a particular feedstock in terms of environmental impact and indirect natural resource costs towards different services and commodities [108]. LCA allows the possibility of comparing different H_2 production approaches and identifying the environmental "hot spot" of the whole process, which helps in development of a sustainable H_2 production process [106,109]. Investigations of the environmental benefits and impacts from a life cycle perspective are scarce. Only a few LCA-studies have been performed specifically on H_2 production. The feedstocks investigated so far are steamed potato peel, wheat straw and sweet sorghum stalks [110,111,112].

In connection with a European research study, HYVOLUTION, the life cycle environmental impacts of pilot production of H_2 through thermophilic fermentation, and photo fermentation of potato peel was compared to production of H_2 from natural gas through steam methane reforming (SMR) [112]. It was demonstrated that the bio-H_2 production had approximately 5.7 times higher environmental impacts (negative impacts on the environment) than a centralized SMR. The processes involved in steam (pretreatment), phosphate buffer (used in photo fermentation) and potassium hydroxide (used in thermophilic fermentation), were the main causes of the environmental impact (98.3%). Recirculation of the sewage reduces the environmental impacts considerably to having only approximately two times more environmental impact than SMR. If instead biomethane were produced for use in the SMR the environmental impact would be reduced to less than 1/3 of the traditional SMR [112]. On the other hand alternative use of the peel would be as animal feed and Djomo et al. showed that the production of bio-H_2 is more beneficial than the use as animal feed by a factor of 2–3 [110]. In a more recent study Djomo and Blumberga investigated potential differences in environmental performance between the three different feedstocks [111]. They performed a "well-to-tank" study i.e., the system boundary is at supplying H_2 to road vehicles meaning that the combustion and transportation of H_2 in the vehicles, was not included. Further, the production of feedstock was excluded as they are considered wastes. Their conclusion is in contrast to the earlier study they find that H_2 produced from any of the feedstock reduced GHG-emissions by approximately 55% compared to SMR and a few percent less for gasoline. When the subsequent use of the remains from the H_2 production were considered as animal feed, an environmental benefit could be observed. The energy ratio calculated was 1.08–1.17, i.e., the energy gain is between 8 and 17%. Though steamed potato peel was slightly better, no significant environmental differences were observed between the feedstocks [111]. The results compare well with those of Manish and Banerjee who investigated the energy balance of H_2 and found an energy ratio of 3.1 (excluding the gas treatment and the compressing) [57].

The conclusion from these studies from an environmental view point is that the production of H_2 for renewable energy production from potato peel could be preferred to using SMR or as direct animal feed due to the lesser environmental impacts. The LCA studies can further be used for identification of the main environmental improvements in the technology development (e.g., recirculation of the sewage and reuse of the remains for animal feed). The LCA of H_2 is very important before taking them into consideration for commercial scale production and policy decisions on H_2 promotion.

6. FUTURE DIRECTIONS AND PERSPECTIVES

One option proposed to lower feedstock costs is to identify microbes that can directly utilize hemicellulose and cellulose [26]. This would eliminate the need for cellulase enzymes and simplify biomass pretreatment. As cellulose is the most abundant biopolymer in the world [141], its bioconversion provides a viable approach to produce renewable H_2 from organic matter. The combined dark fermentation coupling with photo fermentation, or dark fermentation coupling with bioelectrohydrogenesis is a promising H_2 production process from lignocellulosic biomass if the technological barriers can be overcome [12]. Overall, to develop a mature H_2 production technology, bioconversion performance from lignocellulosic biomass need to be further improved in terms of production rates, cost-effectiveness, and system scale-up. Based on the limited number of LCA studies done on H_2 production, it can be assumed that the bioconversion of lignocelluloses-to-H_2 on industrial scale is a feasible option to produce H_2 via biotechnology. However, more in-depth studies need to be carried out to confirm this.

7. CONCLUSIONS

Although considerable progress has been made on H_2 production from lignocellulosic biomass, several challenges remain for its commercial application. Among the various techniques available for H_2 production from lignocellulosic biomass, dark fermentation seems to have an edge over the others and is the closest to commercialization. Photo fermentation is the next best option, though it has to overcome the problems associated with reactor design and operation. Bioelectrochemical H_2 production is still in its infancy and needs much more research and development. The kinetic models for H_2 production provide insights on substrate utilization and factors limiting higher yields. The models will help in scale up studies for validating the proposed data and later on with the experimental data. The few environmental assessment studies performed from a LCA perspective show that H_2 production from lignocellulosic biomass also may be preferable to other renewable energy production pathways. Such studies can furthermore help identifying technological improvement options. The results of LCA studies could also help policy makers in taking decision on policies related to promotion of renewable energy.

AUTHOR CONTRIBUTIONS

Anoop Singh and Deepak Pant originated the manuscript idea, created the outline, wrote parts of the manuscript and did the final editing. Surajbhan Sevda and Ibrahim M. Abu Reesh wrote about the modelling aspect and prepared the final revised draft. Karolien Vanbroekhoven and Dheeraj Rathore contributed to different sections of the manuscript mainly microbial electrolysis and sustainability.

CONFLICTS OF INTEREST

The authors declare no conflict of interest.

REFERENCES

1. Prasad, S.; Singh, A.; Joshi, H.C. Ethanol as an alternative fuel from agricultural, industrial and urban residues. Resour. Conserv. Recycl. 2007, 50, 1–39.

2. Prasad, S.; Singh, A.; Joshi, H.C. Ethanol production from sweet sorghum syrup for utilization as automotive fuel in India. Energy Fuels 2007, 21, 2415–2420.

3. Pant, D.; Van Bogaert, G.; Diels, L.; Vanbroekhoven, K. A review of the substrates used in microbial fuel cells (MFCs) for sustainable energy production. Bioresour. Technol. 2010, 101, 1533–1543.

4. Pant, D.; Singh, A.; van Bogaert, G.; Gallego, Y.A.; Diels, L.; Vanbroekhoven, K. An introduction to the life cycle

assessment (LCA) of bioelectrochemical systems (BES) for sustainable energy and product generation: Relevance and key aspects. Renew. Sustain. Energy Rev. 2011, 15, 1305–1313.

5. Singh, A.; Smyth, B.M.; Murphy, J.D. A biofuel strategy for Ireland with an emphasis on production of biomethane and minimization of land-take. Renew. Sustain. Energy Rev. 2010, 14, 277–288.

6. Singh, A.; Pant, D.; Korres, N.E.; Nizami, A.; Prasad, S.; Murphy, J.D. Key issues in life cycle assessment of ethanol production from lignocellulosic biomass: Challenges and perspectives. Bioresour. Technol. 2010, 101, 5003–5012.

7. Singh, A.; Nigam, P.S.; Murphy, J.D. Renewable fuels from algae: An answer to debatable land based fuels. Bioresour. Technol. 2011, 102, 10–16.

8. Hallenbeck, P.C.; Ghosh, D. Advances in fermentative biohydrogen production: The way forward? Trends Biotechnol. 2009, 27, 287–297.

9. Ball, M.; Wietschel, M. The future of hydrogen—Opportunities and challenges. Int. J. Hydrog. Energy 2009, 34, 615–627.

10. Perera, K.R.J.; Ketheesan, B.; Gadhamshetty, V.; Nirmalakhandan, N. Fermentative biohydrogen production: Evaluation of net energy gain. Int. J. Hydrog. Energy 2010, 35, 12224–12233.

11. Kovacs, K.; Maroti, G.; Rakhely, G. A novel approach for biohydrogen production. Int. J. Hydrog. Energy 2006, 31, 1460–1468.

12. Ren, N.; Wang, A.; Cao, G.; Xu, J.; Gao, L. Bioconversion of lignocellulosic biomass to hydrogen: Potential and challenges. Biotechnol. Adv. 2009, 27, 1051–1060.

13. Kumar, N.; Das, D. Enhancement of hydrogen production by Enterobacter cloacae IIT-BT 08. Process Biochem. 2000, 35, 589–593.

14. Chen, C.C.; Lin, C.Y.; Chang, J.S. Kinetics of hydrogen production with continuous anaerobic cultures utilizing sucrose as the limiting substrate. Appl. Microbiol. Biotechnol. 2001, 57, 56–64.

15. Ginkel, S.V.; Sung, S.; Lay, J.J. Biohydrogen production as a function of pH and substrate concentration. Environ. Sci. Technol. 2001, 35, 4726–4730.

16. Xing, D.; Ren, N.; Li, Q.; Lin, M.; Wang, A.; Zhao, L. Ethanoligenens harbinense gen. nov., sp. nov., isolated from molasses wastewater. Int. J. Syst. Evolut. Microbiol. 2006, 56, 755–760.

17. Wang, L.; Zhou, Q.; Li, F. Avoiding propionic acid accumulation in the anaerobic process for biohydrogen production. Biomass Bioenergy 2006, 30, 177–182.

18. Ren, N.; Cao, G.; Guo, W.; Wang, A.; Zhu, Y.; Liu, B.; Xu, J. Biological hydrogen production from corn stover by moderately thermophile Thermoanaerobacterium thermosaccharolyticum W16. Int. J. Hydrog. Energy 2010, 35, 2708–2712.

19. De Vrije, T.; de Haas, G.; Tan, G.B.; Keijsers, E.R.; Claassen, P.A.M. Pretreatment of Miscanthus for hydrogen production by Thermotoga elfii. Int. J. Hydrog. Energy 2002, 27, 1381–1390.

20. Ntaikou, I.; Gavala, H.N.; Kornaros, M.; Lyberatos, G. Hydrogen production from sugars and sweet sorghum biomass using Ruminococcus albus. Int. J. Hydrog. Energy 2008, 33, 1153–1163.

21. Kapdan, I.K.; Kargi, F. Bio-hydrogen production from waste materials. Enzym. Microb. Technol. 2006, 38, 569–582.

22. Kalinci, Y.; Hepbasli, A.; Dincer, I. Biomass-based hydrogen production: A review and analysis. Int. J. Hydrog. Energy 2009, 34, 8799–8817.

23. Kraemer, J.T.; Bagley, D.M. Improving the yield from fermentative hydrogen production. Biotechnol. Lett. 2007, 29, 685–695.

24. Wang, J.; Wan, W. Factors influencing fermentative hydrogen production: A review. Int. J. Hydrog. Energy 2009, 34, 799–811.

25. Cournac, L.; Sarma, P.M.; Fontecave, M. Biohydrogen: From Basic Concepts to Technology. Int. J. Hydrog. Energy 2010, 35, 10638.

26. Turner, J.; Sverdrup, G.; Mann, M.K.; Maness, P.C.; Kroposki, B.; Ghirardi, M.; Evans, R.J.; Blake, D. Renewable hydrogen production. Int. J. Energy Res. 2008, 32, 379–407.

27. Bartacek, J.; Zabranska, J.; Lens, P.N. Developments and constraints in fermentative hydrogen production. Biofuels Bioprod. Biorefining 2007, 1, 201–214.

28. Nielsen, A.T.; Amandusson, H.; Bjorklund, R.; Dannetun, H.; Ejlertsson, J.; Ekedahl, L.G.; Lundström, I.; Svensson, B.H. Hydrogen production from organic waste. Int. J. Hydrog. Energy 2001, 26, 547–550.

29. Kim, S.H.; Shin, H.S. Effects of base-pretreatment on continuous enriched culture for hydrogen production from food waste. Int. J. Hydrog. Energy 2008, 33, 5266–5274.

30. Kim, D.H.; Kim, S.H.; Shin, H.S. Hydrogen fermentation of food waste without inoculum addition. Enzyme Microb. Technol. 2009, 45, 181–187.

31. Ren, N.; Li, J.; Li, B.; Wang, Y.; Liu, S. Biohydrogen production from molasses by anaerobic fermentation with a pilot-scale bioreactor system. Int. J. Hydrog. Energy 2006, 31, 2147–2157.

32. Li, J.; Li, B.; Zhu, G.; Ren, N.; Bo, L.; He, J. Hydrogen production from diluted molasses by anaerobic hydrogen producing bacteria in an anaerobic baffled reactor (ABR). Int. J. Hydrog. Energy 2007, 32, 3274–3283.

33. Yang, P.; Zhang, R.; Mcgarvey, J.; Benemann, J. Biohydrogen production from cheese processing wastewater by anaerobic fermentation using mixed microbial communities. Int. J. Hydrog. Energy 2007, 32, 4761–4771.

34. Azbar, N.; Çetinkaya Dokgöz, F.T.; Keskin, T.; Korkmaz, K.S.; Syed, H.M. Continuous fermentative hydrogen production from cheese whey wastewater under thermophilic anaerobic conditions. Int. J. Hydrog. Energy 2009, 34, 7441–7447.

35. Lay, J.J. Modeling and optimization of anaerobic digested sludge converting starch to hydrogen. Biotechnol. Bioeng. 2000, 63, 269–278.

36. Ueno, Y.; Haruta, S.; Ishii, M.; Igarashi, Y. Microbial community in anaerobic hydrogen-producing microflora enriched from sludge compost. Appl. Microbiol. Biotechnol. 2001, 57, 555–562.

37. Lin, C.Y.; Cheng, C.H. Fermentative hydrogen production from xylose using anaerobic mixed microflora. Int. J. Hydrog. Energy 2006, 31, 832–840.

38. Hussy, I.; Hawkes, F.R.; Dinsdale, R.; Hawkes, D.L. Continuous fermentative hydrogen production from sucrose and sugarbeet. Int. J. Hydrog. Energy 2005, 30, 471–483.

39. Wang, C.H.; Lin, P.J.; Chang, J.S. Fermentative conversion of sucrose and pineapple waste into hydrogen gas in phosphate-buffered culture seeded with municipal sewage sludge. Process Biochem. 2006, 41, 1353–1358.

40. Fan, K.S.; Kan, N.R.; Lay, J.J. Effect of hydraulic retention time on anaerobic hydrogenesis in CSTR. Bioresour. Technol. 2006, 97, 84–89.

41. Hawkes, F.R.; Dinsdale, R.; Hawkes, D.L.; Hussy, I. Sustainable fermentative hydrogen production: Challenges for process optimisation. Int. J. Hydrog. Energy 2002, 27, 1339–1347.

42. Hawkes, F.R.; Hussy, I.; Kyazze, G.; Dinsdale, R.; Hawkes, D.L. Continuous dark fermentative hydrogen production by mesophilic microflora: Principles and progress. Int. J. Hydrog. Energy 2007, 32, 172–184.

43. Varanasi, J.L.; Roy, S.; Pandit, S.; Das, D. Improvement of energy recovery from cellobiose by thermophillic dark fermentative hydrogen production followed by microbial fuel cell. Int. J. Hydrog. Energy 2015, 40, 8311–8321.

44. Angenent, L.T.; Karim, K.; Al-Dahhan, M.H.; Wrenn, B.A.; Domíguez-Espinosa, R. Production of bioenergy and biochemicals from industrial and agricultural wastewater. Trends Biotechnol. 2004, 22, 477–485.

45. Levin, D. Biohydrogen production: Prospects and limitations to practical application. Int. J. Hydrog. Energy 2004, 29, 173–185.

46. Dipasquale, L.; D'Ippolito, G.; Fontana, A. Capnophilic lactic fermentation and hydrogen synthesis by Thermotoga neapolitana: An unexpected deviation from the dark fermentation model. Int. J. Hydrog. Energy 2014, 39, 4857–4862.

47. Pradhan, N.; Dipasquale, L.; D'Ippolito, G.; Panico, A.; Lens, P.N.L.; Esposito, G.; Fontana, A. Hydrogen production by the thermophilic bacterium thermotoga neapolitana. Int. J. Mol. Sci. 2015, 16, 12578–12600.

48. Basak, N.; Das, D. The prospect of purple non-sulfur (PNS) photosynthetic bacteria for hydrogen production: The present state of the art. World J. Microbiol. Biotechnol. 2007, 23, 31–42.

49. Redwood, M.D.; Paterson-Beedle, M.; MacAskie, L.E. Integrating dark and light bio-hydrogen production

strategies: Towards the hydrogen economy. Rev. Environ. Sci. Biotechnol. 2009, 8, 149–185.

50. Holladay, J.D.; Hu, J.; King, D.L.; Wang, Y. An overview of hydrogen production technologies. Catal. Today 2009, 139, 244–260.

51. Kapdan, I.K.; Kargi, F.; Oztekin, R.; Argun, H. Bio-hydrogen production from acid hydrolyzed wheat starch by photo-fermentation using different Rhodobacter sp. Int. J. Hydrog. Energy 2009, 34, 2201–2207.

52. Keskin, T.; Hallenbeck, P.C. Hydrogen production from sugar industry wastes using single-stage photofermentation. Bioresour. Technol. 2012, 112, 131–136.

53. Nath, K.; Das, D. Improvement of fermentative hydrogen production: Various approaches. Appl. Microbiol. Biotechnol. 2004, 65, 520–529.

54. Liu, B.F.; Ren, N.Q.; Xie, G.J.; Ding, J.; Guo, W.Q.; Xing, D.F. Enhanced bio-hydrogen production by the combination of dark- and photo-fermentation in batch culture. Bioresour. Technol. 2010, 101, 5325–5329.

55. Logan, B.E. Microbial Fuel Cell; Wiley & Sons, Inc.: Hoboken, NJ, USA, 2008.

56. Rozendal, R.; Hamelers, H.; Euverink, G.; Metz, S.; Buisman, C. Principle and perspectives of hydrogen production through biocatalyzed electrolysis. Int. J. Hydrog. Energy 2006, 31, 1632–1640.

57. Manish, S.; Banerjee, R. Comparison of biohydrogen production processes. Int. J. Hydrog. Energy 2008, 33, 279–286.

58. Cusick, R.D.; Bryan, B.; Parker, D.S.; Merrill, M.D.; Mehanna, M.; Kiely, P.D.; Liu, G.; Logan, B.E. Performance of a pilot-scale continuous flow microbial electrolysis cell fed winery

wastewater. Appl. Microbiol. Biotechnol. 2011, 89, 2053–2063.

59. Kadier, A.; Simayi, Y.; Kalil, M.S.; Abdeshahian, P.; Hamid, A.A. A review of the substrates used in microbial electrolysis cells (MECs) for producing sustainable and clean hydrogen gas. Renew. Energy 2014, 71, 466–472.

60. Ren, L.; Siegert, M.; Ivanov, I.; Pisciotta, J.M.; Logan, B.E. Treatability studies on different refinery wastewater samples using high-throughput microbial electrolysis cells (MECs). Bioresour. Technol. 2013, 136, 322–328.

61. Hu, H.; Fan, Y.; Liu, H. Hydrogen production in single-chamber tubular microbial electrolysis cells using non-precious-metal catalysts. Int. J. Hydrog. Energy 2009, 34, 8535–8542.

62. Lu, L.; Xing, D.; Ren, N.; Logan, B.E. Syntrophic interactions drive the hydrogen production from glucose at low temperature in microbial electrolysis cells. Bioresour. Technol. 2012, 124, 68–76.

63. Lu, L.; Ren, N.; Xing, D.; Logan, B.E. Hydrogen production with effluent from an ethanol-H2-coproducing fermentation reactor using a single-chamber microbial electrolysis cell. Biosens. Bioelectron. 2009, 24, 3055–3060.

64. Call, D.; Logan, B.E. Hydrogen production in a single chamber microbial electrolysis cell lacking a membrane. Environ. Sci. Technol. 2008, 42, 3401–3406.

65. Call, D.F.; Merrill, M.D.; Logan, B.E. High surface area stainless steel brushes as cathodes in microbial electrolysis cells. Environ. Sci. Technol. 2009, 43, 2179–2183.

66. Selembo, P.A.; Perez, J.M.; Lloyd, W.A.; Logan, B.E. High hydrogen production from glycerol or glucose by electrohydrogenesis using microbial electrolysis cells. Int. J. Hydrog. Energy 2009, 34, 5373–5381.

67. Wang, X.; Cheng, S.; Feng, Y.; Merrill, M.D.; Saito, T.; Logan, B.E. Use of carbon mesh anodes and the effect of different pretreatment methods on power production in microbial fuel cells. Environ. Sci. Technol. 2009, 43, 6870–6874.

68. Wagner, R.C.; Regan, J.M.; Oh, S.-E.; Zuo, Y.; Logan, B.E. Hydrogen and methane production from swine wastewater using microbial electrolysis cells. Water Res. 2009, 43, 1480–1488.

69. Pisciotta, J.M.; Zaybak, Z.; Call, D.F.; Nam, J.Y.; Logan, B.E. Enrichment of microbial electrolysis cell biocathodes from sediment microbial fuel cell bioanodes. Appl. Environ. Microbiol. 2012, 78, 5212–5219.

70. Xiao, L.; Wen, Z.; Ci, S.; Chen, J.; He, Z. Carbon/iron-based nanorod catalysts for hydrogen production in microbial electrolysis cells. Nano Energy 2012, 1, 751–756.

71. Guo, K.; Tang, X.; Du, Z.; Li, H. Hydrogen production from acetate in a cathode-on-top single-chamber microbial electrolysis cell with a mipor cathode. Biochem. Eng. J. 2010, 51, 48–52.

72. Hu, H.; Fan, Y.; Liu, H. Hydrogen production using single-chamber membrane-free microbial electrolysis cells. Water Res. 2008, 42, 4172–4178.

73. Van Groenestijn, J.W.; Geelhoed, J.S.; Goorissen, H.P.; Meesters, K.P.; Stams, A.J.; Claassen, P.A. Performance and population analysis of a non-sterile trickle bed reactor inoculated with caldicellulosiruptor saccharolyticus, a thermophilic hydrogen producer. Biotechnol. Bioeng. 2009, 102, 1361–1367.

74. Kleerebezem, R.; van Loosdrecht, M.C. Mixed culture biotechnology for bioenergy production. Curr. Opin. Biotechnol. 2007, 18, 207–212.

75. Kengen, S.W.M.; Goorissen, H.P.; Verhaart, M.; Stams, A.J.M.; van Niel, E.W.J.; Claassen, P.A.M. Biological Hydrogen

Production by Anaerobic Microorganisms. In Biofuels; John Wiley & Sons, Ltd.: Chichester, UK, 2009; pp. 197–221.

76. Kotsopoulos, T.A.; Zeng, R.J.; Angelidaki, I. Biohydrogen production in granular up-flow anaerobic sludge blanket (UASB) reactors with mixed cultures under hyper-thermophilic temperature (70 C). Biotechnol. Bioeng. 2006, 94, 296–302.

77. De Vrije, T.; Bakker, R.R.; Budde, M.A.; Lai, M.H.; Mars, A.E.; Claassen, P.A. Efficient hydrogen production from the lignocellulosic energy crop Miscanthus by the extreme thermophilic bacteria Caldicellulosiruptor saccharolyticus and Thermotoga neapolitana. Biotechnol. Biofuels 2009, 2, 12.

78. Van Niel, E.W.; Budde, M.A.; De Haas, G.; Van der Wal, F.J.; Claassen, P.A.; Stams, A.J. Distinctive properties of high hydrogen producing extreme thermophiles, Caldicellulosiruptor saccharolyticus and Thermotoga elfii. Int. J. Hydrog. Energy 2002, 27, 1391–1398.

79. Thauer, R.K.; Jungermann, K.; Decker, K. Energy conservation in chemotrophic anaerobic bacteria. Bacteriol. Rev. 1977, 41, 100–180.

80. Vanginkel, S.; Oh, S.; Logan, B. Biohydrogen gas production from food processing and domestic wastewaters. Int. J. Hydrog. Energy 2005, 30, 1535–1542.

81. Liu, D.; Liu, D.; Zeng, R.J.; Angelidaki, I. Hydrogen and methane production from household solid waste in the two-stage fermentation process. Water Res. 2006, 40, 2230–2236.

82. Oh, S.-E.; van Ginkel, S.; Logan, B.E. The relative effectiveness of pH control and heat treatment for enhancing biohydrogen gas production. Environ. Sci. Technol. 2003, 37, 5186–5190.

83. Liang, T.M.; Cheng, S.S.; Wu, K.L. Behavioral study on hydrogen fermentation reactor installed with silicone rubber membrane. Int. J. Hydrog. Energy 2002, 27, 1157–1165.

84. Van Groenestijn, J.W.; Hazewinkel, J.H.; Nienoord, M.; Bussmann, P.J. Energy aspects of biological hydrogen production in high rate bioreactors operated in the thermophilic temperature range. Int. J. Hydrog. Energy 2002, 27, 1141–1147.

85. Kim, D.H.; Han, S.K.; Kim, S.H.; Shin, H.S. Effect of gas sparging on continuous fermentative hydrogen production. Int. J. Hydrog. Energy 2006, 31, 2158–2169.

86. Koku, H. Aspects of the metabolism of hydrogen production by Rhodobacter sphaeroides. Int. J. Hydrog. Energy 2002, 27, 1315–1329.

87. Beckers, L.; Masset, J.; Hamilton, C.; Delvigne, F.; Toye, D.; Crine, M.; Thonart, P.; Hiligsmann, S. Investigation of the links between mass transfer conditions, dissolved hydrogen concentration and biohydrogen production by the pure strain Clostridium butyricum CWBI1009. Biochem. Eng. J. 2015, 98, 18–28.

88. Mizuno, O.; Dinsdale, R.; Hawkes, F.R.; Hawkes, D.L.; Noike, T. Enhancement of hydrogen production from glucose by nitrogen gas sparging. Bioresour. Technol. 2000, 73, 59–65.

89. Chou, C.; Wang, C.; Huang, C.; Lay, J. Pilot study of the influence of stirring and pH on anaerobes converting high-solid organic wastes to hydrogen. Int. J. Hydrog. Energy 2008, 33, 1550–1558.

90. Fontes Lima, D.M.; Zaiat, M. The influence of the degree of back-mixing on hydrogen production in an anaerobic fixed-bed reactor. Int. J. Hydrog. Energy 2012, 37, 9630–9635.

91. Montiel-Corona, V.; Revah, S.; Morales, M. Hydrogen production by an enriched photoheterotrophic culture using dark fermentation effluent as substrate: Effect of flushing method, bicarbonate addition, and outdoor–indoor conditions. Int. J. Hydrog. Energy 2015, 40, 9096–9105.

92. Clark, I.C.; Zhang, R.H.; Upadhyaya, S.K. The effect of low pressure and mixing on biological hydrogen production via anaerobic fermentation. Int. J. Hydrog. Energy 2012, 37, 11504–11513.

93. Kim, D.-H.; Shin, H.-S.; Kim, S.-H. Enhanced H2 fermentation of organic waste by CO2 sparging. Int. J. Hydrog. Energy 2012, 37, 15563–15568.

94. Gómez, X.; Morán, A.; Cuetos, M.J.; Sánchez, M.E. The production of hydrogen by dark fermentation of municipal solid wastes and slaughterhouse waste: A two-phase process. J. Power Sources 2006, 157, 727–732.

95. Clauwaert, P.; Verstraete, W. Methanogenesis in membraneless microbial electrolysis cells. Appl. Microbiol. Biotechnol. 2009, 82, 829–836.

96. Cheng, S.; Xing, D.; Call, D.F.; Logan, B.E. Direct biological conversion of electrical current into methane by electromethanogenesis. Environ. Sci. Technol. 2009, 43, 3953–3958.

97. Pant, D.; van Bogaert, G.; de Smet, M.; Diels, L.; Vanbroekhoven, K. Use of novel permeable membrane and air cathodes in acetate microbial fuel cells. Electrochimica Acta 2010, 55, 7710–7716.

98. Selembo, P.A.; Merrill, M.D.; Logan, B.E. The use of stainless steel and nickel alloys as low-cost cathodes in microbial electrolysis cells. J. Power Sour. 2009, 190, 271–278.

99. Maeda, T.; Sanchez-Torres, V.; Wood, T.K. Metabolic engineering to enhance bacterial hydrogen production. Microb. Biotechnol. 2008, 1, 30–39.

100. Morimoto, K.; Kimura, T.; Sakka, K.; Ohmiya, K. Overexpression of a hydrogenase gene in Clostridium paraputrificum to enhance hydrogen gas production. FEMS Microbiol. Lett. 2005, 246, 229–234.

101. Vardar-Schara, G.; Maeda, T.; Wood, T.K. Metabolically engineered bacteria for producing hydrogen via fermentation. Microb. Biotechnol. 2008, 1, 107–125.

102. Hallenbeck, P.C.; Ghosh, D. Improvements in fermentative biological hydrogen production through metabolic engineering. J. Environ. Manag. 2012, 95, S360–S364.

103. Ryu, M.-H.; Hull, N.C.; Gomelsky, M. Metabolic engineering of Rhodobacter sphaeroides for improved hydrogen production. Int. J. Hydrog. Energy 2014, 39, 6384–6390.

104. Wang, J.; Wan, W. Kinetic models for fermentative hydrogen production: A review. Int. J. Hydrog. Energy 2009, 34, 3313–3323.

105. Dincer, I. Hydrogen and Fuel Cell Technologies for Sustainable Future. Jordan J. Mech. Ind. Eng. 2008, 2, 1–14.

106. Rathore, D.; Singh, A. Biohydrogen production from microalgae. In Biofuels Technologies Recent Developments; Gupta, V., Tuohy, M., Eds.; Springer: Berlin/Heidelberg, Germany, 2013; pp. 317–333.

107. Kwak, H.Y.; Lee, H.S.; Jung, J.Y.; Jeon, J.S.; Park, D.R. Exergetic and thermoeconomic analysis of a 200-kW phosphoric acid fuel cell plant. Fuel 2004, 83, 2087–2094.

108. Rubio Rodríguez, M.A.; de Ruyck, J.; Díaz, P.R.; Verma, V.K.; Bram, S. An LCA based indicator for evaluation of alternative energy routes. Appl. Energy 2011, 88, 630–635.

109. Romagnoli, F.; Blumberga, D.; Pilicka, I. Life cycle assessment of biohydrogen production in photosynthetic processes. Int. J. Hydrog. Energy 2011, 36, 7866–7871.

110. Djomo, S.N.; Humbert, S. Dagnija Blumberga Life cycle assessment of hydrogen produced from potato steam peels. Int. J. Hydrog. Energy 2008, 33, 3067–3072.

111. Djomo, S.N.; Blumberga, D. Comparative life cycle assessment of three biohydrogen pathways. Bioresour. Technol. 2011, 102, 2684–2694.

112. Ochs, D.; Wukovits, W.; Ahrer, W. Life cycle inventory analysis of biological hydrogen production by thermophilic and photo fermentation of potato steam peels (PSP). J. Clean. Prod. 2010, 18, S88–S94.

113. Chen, S.D.; Lo, Y.C.; Lee, K.S.; Huang, T.I.; Chang, J.S. Sequencing batch reactor enhances bacterial hydrolysis of starch promoting continuous bio-hydrogen production from starch feedstock. Int. J. Hydrog. Energy 2009, 34, 8549–8557.

114. Seifert, K.; Waligorska, M.; Wojtowski, M.; Laniecki, M. Hydrogen generation from glycerol in batch fermentation process. Int. J. Hydrog. Energy 2009, 34, 3671–3678.

115. Lin, C.N.; Wu, S.Y.; Chang, J.S.; Chang, J.S. Biohydrogen production in a three-phase fluidized bed bioreactor using sewage sludge immobilized by ethylene-vinyl acetate copolymer. Bioresour. Technol. 2009, 100, 3298–3301.

116. Yasin Nazlina, H.M.; Aini, R.; Ismail, F.; Zulkhairi, M.; Hassan, M.A. Effect of different temperature, initial pH and substrate composition on biohydrogen production from food waste in batch fermentation. Asian J. Biotechnol. 2009, 1, 42–50.

117. Zhu, J.; Li, Y.; Wu, X.; Miller, C.; Chen, P.; Ruan, R. Swine manure fermentation for hydrogen production. Bioresour. Technol. 2009, 100, 5472–5477.

118. Gadhamshetty, V.; Johnson, D.C.; Nirmalakhandan, N.; Smith, G.B.; Deng, S. Feasibility of biohydrogen production at low temperatures in unbuffered reactors. Int. J. Hydrog. Energy 2009, 34, 1233–1243.

119. Ghosh, D.; Hallenbeck, P.C. Fermentative hydrogen yields from different sugars by batch cultures of metabolically engineered Escherichia coli DJT135. Int. J. Hydrog. Energy 2009, 34, 7979–7982.

120. Kargi, F.; Pamukoglu, M.Y. Dark fermentation of ground wheat starch for bio-hydrogen production by fed-batch operation. Int. J. Hydrog. Energy 2009, 34, 2940–2946.

121. Han, W.; Ye, M.; Zhu, A.J.; Zhao, H.T.; Li, Y.F. Batch dark fermentation from enzymatic hydrolyzed food waste for hydrogen production. Bioresour. Technol. 2015, 191, 24–29.

122. Chookaew, T.; O-Thong, S.; Prasertsan, P. Biohydrogen production from crude glycerol by two stage of dark and photo fermentation. Int. J. Hydrog. Energy 2015, 40, 7433–7438.

123. Wicher, E.; Seifert, K.; Zagrodnik, R.; Pietrzyk, B.; Laniecki, M. Hydrogen gas production from distillery wastewater by dark fermentation. Int. J. Hydrog. Energy 2013, 38, 7767–7773.

124. Moreno, R.; Escapa, A.; Cara, J.; Carracedo, B.; Gómez, X. A two-stage process for hydrogen production from cheese whey: Integration of dark fermentation and biocatalyzed electrolysis. Int. J. Hydrog. Energy 2015, 40, 168–175.

125. Su, H.; Cheng, J.; Zhou, J.; Song, W.; Cen, K. Hydrogen production from water hyacinth through dark- and photo-fermentation. Int. J. Hydrog. Energy 2010, 35, 8929–8937.

126. Argun, H.; Kargi, F. Effects of sludge pre-treatment method on bio-hydrogen production by dark fermentation of waste ground wheat. Int. J. Hydrog. Energy 2009, 34, 8543–8548.

127. Laurinavichene, T.V.; Belokopytov, B.F.; Laurinavichius, K.S.; Tekucheva, D.N.; Seibert, M.; Tsygankov, A.A. Towards the integration of dark- and photo-fermentative waste treatment. 3. Potato as substrate for sequential dark fermentation and light-driven H2 production. Int. J. Hydrog. Energy 2010, 35, 8536–8543.

128. Ozmihci, S.; Kargi, F. Effects of starch loading rate on performance of combined fed-batch fermentation of ground wheat for bio-hydrogen production. Int. J. Hydrog. Energy 2010, 35, 1106–1111.

129. Avcioglu, S.G.; Ozgur, E.; Eroglu, I.; Yucel, M.; Gunduz, U. Biohydrogen production in an outdoor panel photobioreactor on dark fermentation effluent of molasses. Int. J. Hydrog. Energy 2011, 36, 11360–11368.

130. Zhu, Z.; Shi, J.; Zhou, Z.; Hu, F.; Bao, J. Photo-fermentation of Rhodobacter sphaeroides for hydrogen production using lignocellulose-derived organic acids. Process Biochem. 2010, 45, 1894–1898.

131. Francou, N.; Vignais, P.M. Hydrogen production by Rhodopseudomonas capsulata cells entrapped in carrageenan beads. Biotechnol. Lett. 1984, 6, 639–644.

132. Taguchi, F.; Mizukami, N.; Hasegawa, K.; Saito-Taki, T. Microbial conversion of arabinose and xylose to hydrogen by a newly isolated clostridium sp. No. 2. Can. J. Microbiol. 1994, 40, 228–233.

133. Susanti, R.F.; Dianningrum, L.W.; Yum, T.; Kim, Y.; Lee, B.G.; Kim, J. High-yield hydrogen production from glucose by supercritical water gasification without added catalyst. Int. J. Hydrog. Energy 2012, 37, 11677–11690.

134. Qinming, Z.; Shuzhong, W.; Liang, W.; Donghai, X. Catalytic Hydrogen Production from Municipal Sludge in Supercritical Water with Partial Oxidation. Chall. Power Eng. Environ. 2007, 1, 1252–1255.

135. Frusteri, F.; Freni, S.; Chiodo, V.; Spadaro, L.; Bonura, G.; Cavallaro, S. Potassium improved stability of Ni/MgO in the steam reforming of ethanol for the production of hydrogen for MCFC. J. Power Sour. 2004, 132, 139–144.

136. Aiello, R.; Fiscus, J.E.; Loye, H.; Amiridis, M.D. Hydrogen production via the direct cracking of methane over Ni/SiO2: Catalyst deactivation and regeneration. Appl. Catal. A Gen. 2000, 192, 227–234.

137. Deng, W.; Jiang, H.; Wu, Y.; Fan, H.; Ji, J. Hydrogen production from biomass pyrolysis in molten alkali. AASRI Procedia 2012, 3, 217–223.

138. Sivasubramanian, P.; Ramasamy, R.P.; Freire, F.J.; Holland, C.E.; Weidner, J.W. Electrochemical hydrogen production from thermochemical cycles using a proton exchange membrane electrolyzer. Int. J. Hydrog. Energy 2007, 32, 463–468.

139. Kim, J.; Kwon, D.; Kim, K.; Hoffmann, M.R. Electrochemical Production of Hydrogen Coupled with the Oxidation of Arsenite. Environ. Sci. Technol. 2014, 48, 2059–2066.

140. Karn, R.K.; Srivastava, O.N. On the synthesis and photochemical studies of nanostructured TiO 1 and TiO 1 admixed VO 1 photoelectrodes in regard to hydrogen production through photoelectrolysis. Int. J. Hydrog. Energy 1999, 24, 965–971.

141. Perlack, R.D.; Wright, L.L.; Turhollow, A.F.; Graham, R.L.; Stokes, B.J.; Erbach, D.C. Biomass as Feedstock For a Bioenergy and Bioproducts Industry: The Technical Feasibility of a Billion-Ton Annual Supply; U.S. Department of Energy: Oak Ridge, TN, USA, 2005.

6
CHAPTER

EMISSIONS OF DIESEL - VEGETABLE OILS MIXTURES

Charalampos Arapatsakos[1]

[1] Department of Production and Management Engineering, Democritus University of Thrace, Xanthi , Greece

1. INTRODUCTION

The industrialization of society, the introduction of motorized vehicles and the explosion of the population are factors contributing toward the growing air pollution problem. Moreover, the exhaust from burning fuels in automobiles, homes and industries is a major source of pollution in the air. Apart from the anthropogenic sources of air pollution there are natural sources as well. Natural sources related to dust from natural source, usually large areas of land with little or no vegetation, the smoke and carbon monoxide from wildfires, volcanic activity etc. Air pollution not only affects the air we breathe, but it also impacts the land and the water. The human health effects of poor air quality are far reaching, but principally affect the body's respiratory system and the cardiovascular system. The human health effects caused by air pollution may range from subtle biochemical and physiological changes to difficulty breathing. It can also cause deaths, aggravated asthma, bronchitis, emphysema, lung and heart diseases to human beings. There are several many types of air pollutant [1,2]. These include smog, acid rain, the greenhouse effect and holes in the ozone layer.

The atmospheric conditions such as the wind, rain, stability affect the transportation of the air pollutant [3,4]. Furthermore, depending on the geographical location temperature, wind and weather factors, pollution is dispersed differently [5,6]. For instance, the wind and rain may effectively dilute pollution to relatively safe concentrations despite a fairly high rate of emissions. In contrast when atmospheric conditions are stable relatively low emissions can cause buildup of pollution to hazardous levels.

The quality of fuel affects diesel engine emissions (HC, CO, NOx and particulate emissions) very strongly. The fuel that is used in diesel engines is a mixture of hydrocarbons and its boiling temperature is approximately 170°C to 360°C [4]. Diesel fuel emissions composition and characteristics depend on mixture formation and combustion. In order to compare the quality of fuels the following criteria are tested: ketene rating, density, viscosity, boiling characteristics, aromatics content and sylph content. For environmental compatibility, the fuel must have low density, low content of aromatic compounds, low sylph content and high ketene rating [6,7,8].

One of the most important and renewable sources of energy is biomass. Biomass as a renewable source of energy refers to living and recently dead biological material that can be used as fuel or for industrial production. Some examples of biomass fuels are wood, crops, manure and some garbage. Biomass is a renewable energy source due to photosynthesis. Concretely, with the photosynthesis is committed the solar energy and is changed in chemical (energy). At the combustion of biomass the committed solar energy is changed in thermo while the dioxide of coal (CO_2) returns in the atmosphere, while the inorganic elements that are contained in the ash, enrich the soil with nutritious elements. Nowadays, the use of biomass, covers approximately 4% of the total energy which is consumed in USA and 45% of the renewable sources of energy [9,10,11]. The most common source of biomass is the wood. For thousands of years people have burned wood for heating and cooking. Another source of biomass is our garbage that comes from plant or animal products. Moreover, various materials of plant origin, as agricultural remains (e.g. straw), material of animal origin, remains from veterinary surgeon units as well as remains of fishery and their sub products, urban waste etc. Wood waste or garbage can be burned to produce steam for making electricity or to provide heat to industries and homes. Biomass can be used for the production of liquid fuel (called biofuel) which is used for the transportation to many countries of Europe, USA etc. [12,13,14]. Bio-diesel is also produced from oily plants (soya, sunflower) animal greases, products of carcasses, and used oils. Some of biomass advantages which make it an attractive source of energy are the following:

1. Reduction of air pollutants. The combustion of biomass has null balance of dioxide of coal (CO_2,) does not contribute in the

phenomenon of green house, because the quantities of dioxide of coal (CO_2,) that are released at the combustion of biomass are committed again by the plants for the creation of biomass.

2. Zero existence of sulphur in biomass contributes considerably in the restriction of emissions of dioxide of sulphur (SO_2,) that is in charge of the acid rain.
3. Reduction of dependence from imported fuels, improvement of commercial balance, in the guaranty of energy supply and in the saving of exchange.
4. Sources are commonly available.
5. Sources are locally produced, consequently it increases the occupation to the agriculture places with the use of alternatives cultures (several kinds of cane, sorghum), as well as the creation of alternative markets for the traditional cultures (sunflower etc.) and withholding of population in their hearths.
6. Increase of Biomass production can often mean the restoration of waste land.

Biofuels are liquid or gas fuels which are produced from the biomass. Biomass can replace the conventional mineral fuels, totally or partial in the engines [15].

The major issue is how a four-stroke diesel engine behaves on the side of pollutants and operation, when it uses mixed fuel of diesel – vegetable oils.

2. INSTRUMENTATION AND EXPERIMENTAL RESULTS

In the experiment stage has been used directly used vegetable oil (used sunflower oil that emanated from cooking) in the mixture of diesel in to a four – stroke diesel engine. Specifically it has been used diesel, mixture diesel-5% used vegetable oil (u5), diesel-10 used vegetable oil (u10), diesel-20% used vegetable oil (u20), diesel-30% used vegetable oil (u30), diesel-40% used vegetable oil (u40), diesel-50% used vegetable oil (u50) in a four-stroke diesel air-cooled engine named Ruggerini type RD-80, volume 377cc, and power 8.2hp/3000rpm, who was connected with a pump of water centrifugal. Measurements were made when the engine was function on 1000, 1500, 2000 and 2500rpm.

During the experiments, it has been counted:

- The percent of CO
- The ppm of HC
- The ppm of NO
- The percent of smoke

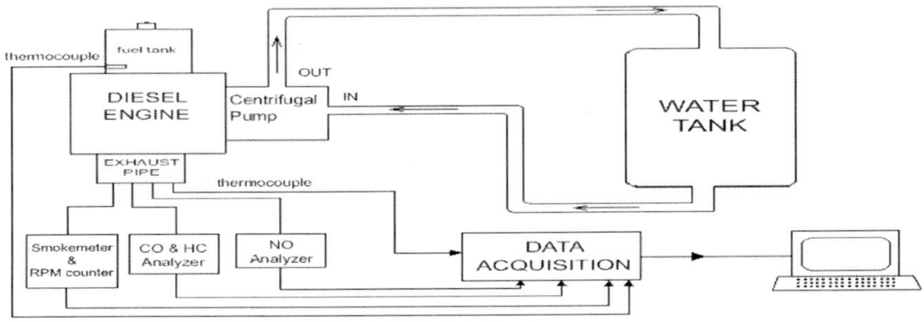

Figure 1. Experimental Layout

The measurement of rounds/min of the engine was made by a portable tachometer (Digital photo/contact tachometer) named LTLutron DT-2236. Smoke was measured by a specifically measurement device named SMOKE MODULE EXHAUST GAS ANALYSER MOD 9010/M, which has been connected to a PC unit. The CO and HC emissions have been measured by HORIBA Analyzer MEXA-324 GE. The NO emissions were measured by a Single GAS Analyser SGA92-NO.

2.1. Used vegetable oil

The experimental results are shown at the following tables and figures [16]:

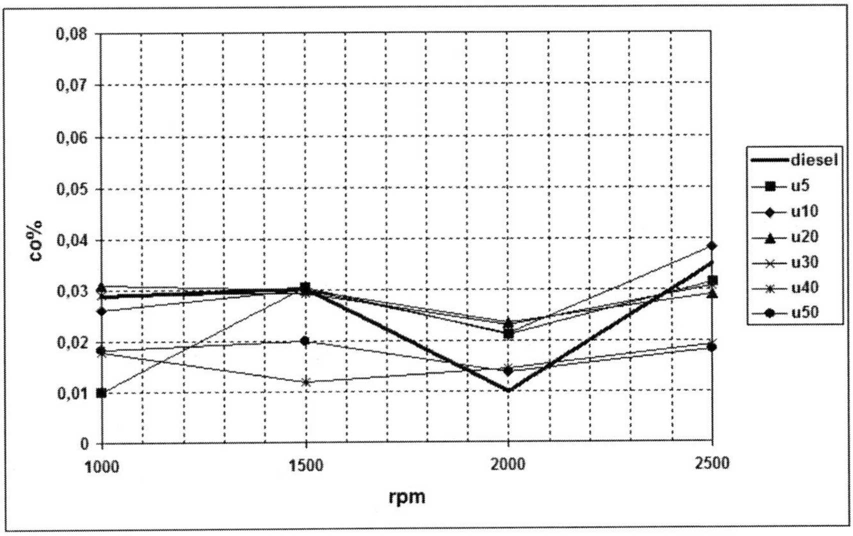

Figure 2. The CO variation on different rpm regarding to the mixture

Table 1. The CO average value variation on different rpm regarding to the mixture

rpm	CO %						
	diesel	u5	u10	u20	u30	u40	u50
1000	0,02898	0,01000	0,026081	0,030985	0,029143	0,017823	0,018223
1500	0,03039	0,03059	0,030043	0,029979	0,029310	0,011818	0,019767
2000	0,01000	0,02108	0,021379	0,023500	0,023059	0,014483	0,013624
2500	0,03508	0,03145	0,038315	0,029120	0,030713	0,019111	0,018298

Table 2. The HC average value variation on different rpm regarding to the mixture

rpm	HC (ppm)						
	diesel	u5	u10	u20	u30	u40	u50
1000	2,535343	8,844156	5,653105	5,246253	5,124364	2,147903	2,974304
1500	13,31714	24,99127	12,87527	13,15385	9,358621	2,934461	6,714588
2000	7,131223	8,326797	12,67026	9,195652	13,79747	5,267241	4,936681
2500	10,96128	16,63420	17,30454	16,94635	6,706013	6,598698	6,759574

Table 3. The NO average value variation on different rpm regarding to the mixture

rpm	NO (ppm)						
	diesel	u5	u10	u20	u30	u40	u50
1000	518,210	771,001	696,827	495,603	380,361	349,140	207,760
1500	739,366	754,126	913,037	771,607	723,381	872,06	582,908
2000	762,155	834,334	520,485	760,936	839,268	928,337	720,505
2500	795,461	946,349	518,287	710,402	864,585	674,432	847,835

Table 4. The % smoke average value variation on different rpm regarding to the mixture

rpm	% smoke						
	diesel	u5	u10	u20	u30	u40	u50
1000	3,262370	4,870779	5,966167	16,43362	12,26745	15,7298	11,32741
1500	7,100651	8,174236	5,768602	7,652778	5,56423	9,206977	13,05011
2000	5,688865	7,619826	4,704957	6,151304	4,948101	4,351724	9,59869
2500	29,00617	23,21970	25,67279	16,86674	14,59399	17,48286	15,87915

From figure 2 it is clear that the more constant behaviour appears in the mixture u40, while the best behaviour is appears in the case diesel/1500rpm. From figure 3 it can be noticed the biggest reduction of HC regarding to diesel in case of mixture u40. From figure 4 it can be noticed the biggest reduction of NO regarding to diesel in the case of mixture u40. From figure 5 it can be seen the biggest reduction for u40 until the case u40/1000rpm. From the above figures it is clear that the use of different mixtures can constitute changes to CO, HC, NO and smoke too. It is also important the fact that there was no changes in the rounds of the engine, as well as in the supply of water at the use of mixtures. Finally as far as the consumption is concerned, did not observed changes with the use of different mixtures.

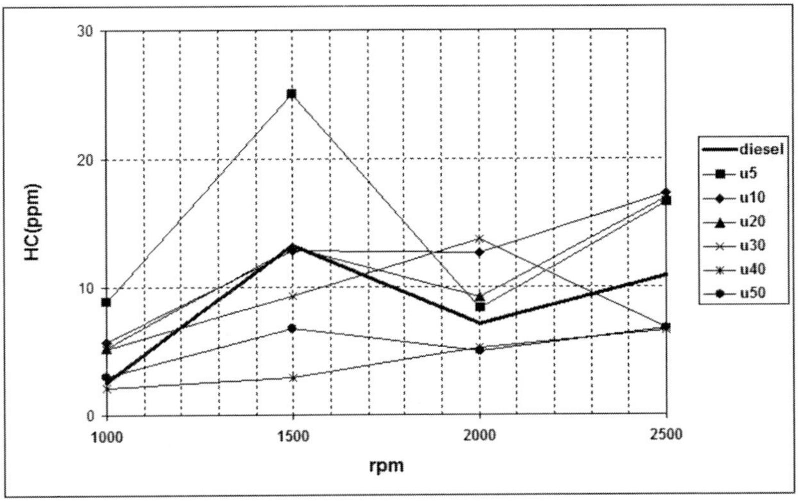

Figure 3. The HC variation on different rpm regarding to the mixture

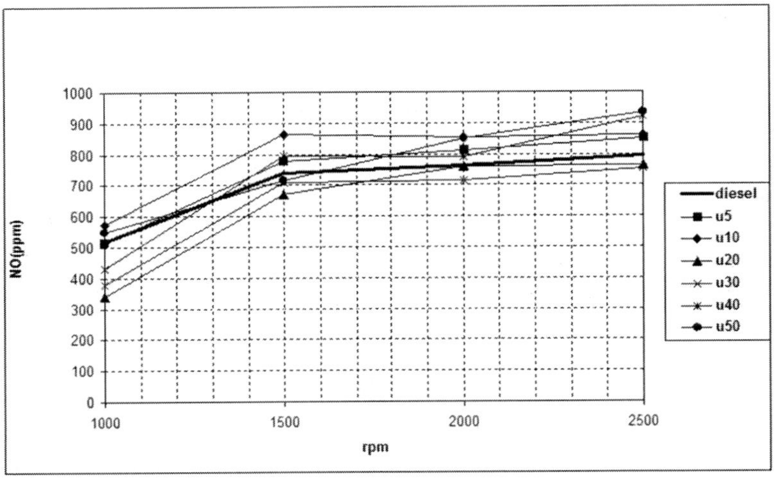

Figure 4. The NO variation on different rpm regarding to the mixture

The use of mixtures of diesel-vegetable oil has as result change of gas emissions with better behaviour in the mixture u40. It is important, that is not presented reduction of power of engine from the combustion of the mixtures.

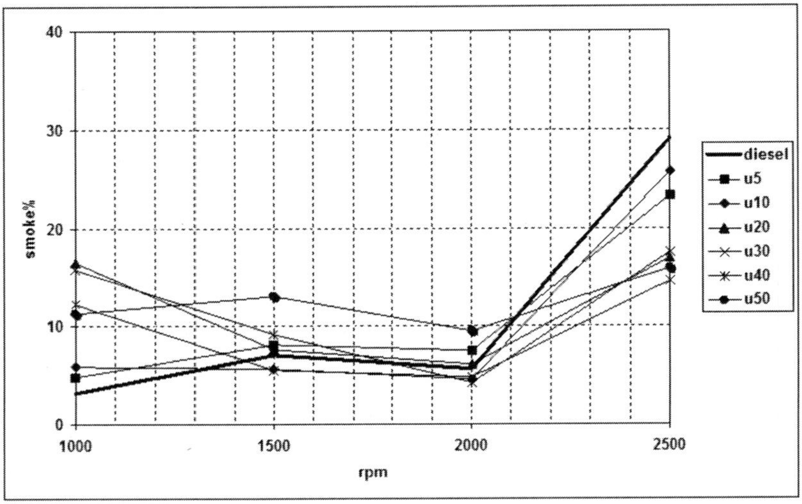

Figure 5. The smoke variation on different rpm regarding to the mixture

2.2. Maize oil

In the experiment stage has been used directly maize oil in the mixture of diesel in to a four – stroke diesel engine. Specifically it has been used diesel, mixture diesel-5% maize oil (k5), diesel-10% maize oil (k10), diesel-20% maize oil (k20), diesel-30% maize oil (k30), diesel-40% maize oil (k40), diesel-50% maize oil (k50) in a four-stroke diesel engine [17]:

Table 5. The CO average value variation on different rpm regarding to the mixture

rpm	CO %						
	diesel	k5	k10	k20	k30	k40	k50
1000	0,0289	0,0310	0,0309	0,0309	0,0319	0,0397	0,0345
1500	0,0303	0,0302	0,0304	0,0311	0,0345	0,0211	0,0288
2000	0,01	0,0280	0,0232	0,0284	0,0274	0,0281	0,0219
2500	0,0350	0,0244	0,0317	0,0296	0,0324	0,0305	0,0292

Table 6. The HC average value variation on different rpm regarding to the mixture

rpm	HC (ppm)						
	diesel	k5	k10	k20	k30	k40	k50
1000	2,535	14,937	6,244	10,326	3,406	5,358	9,167
1500	13,31	21,485	9,236	17,997	14,718	0,449	17,197
2000	7,131	3,184	13,970	15,965	8,402	8,502	12,913
2500	10,961	16,347	18,884	23,556	30,551	7,451	17,712

Table 7. The NO average value variation on different rpm regarding to the mixture

rpm	NO (ppm)						
	diesel	k5	k10	k20	k30	k40	k50
1000	518,210	771,001	696,827	495,603	380,361	349,140	207,760
1500	739,366	754,126	913,037	771,607	723,381	872,06	582,908
2000	762,155	834,334	520,485	760,936	839,268	928,337	720,505
2500	795,461	946,349	518,287	710,402	864,585	674,432	847,835

Table 8. The % smoke average value variation on different rpm regarding to the mixture

rpm	% smoke						
	diesel	k5	k10	k20	k30	k40	k50
1000	3,262	12,722	7,301	7,488	16,623	7,200	26,232
1500	7,100	10,924	5,487	6,547	14,850	12,141	24,035
2000	5,688	18,679	4,001	6,588	9,936	14,071	18,884
2500	29,006	28,282	21,848	15,730	17,579	13,438	14,265

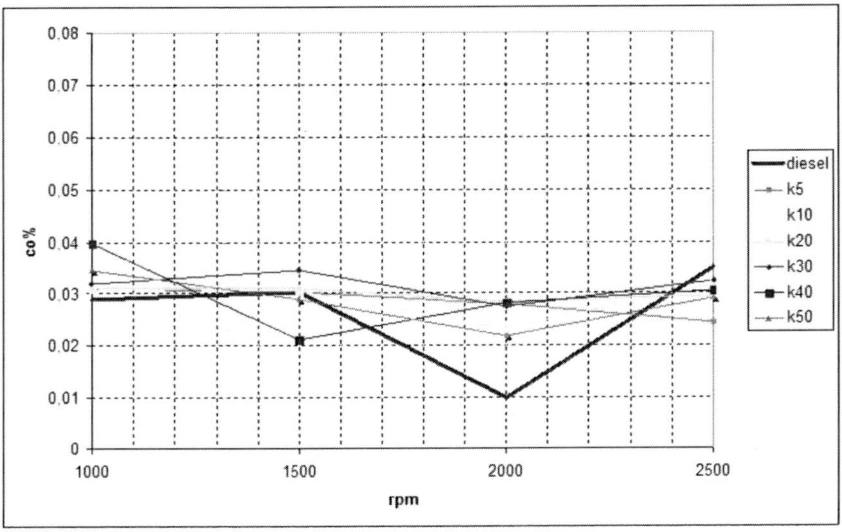

Figure 6. The CO variation on different rpm regarding to the mixture

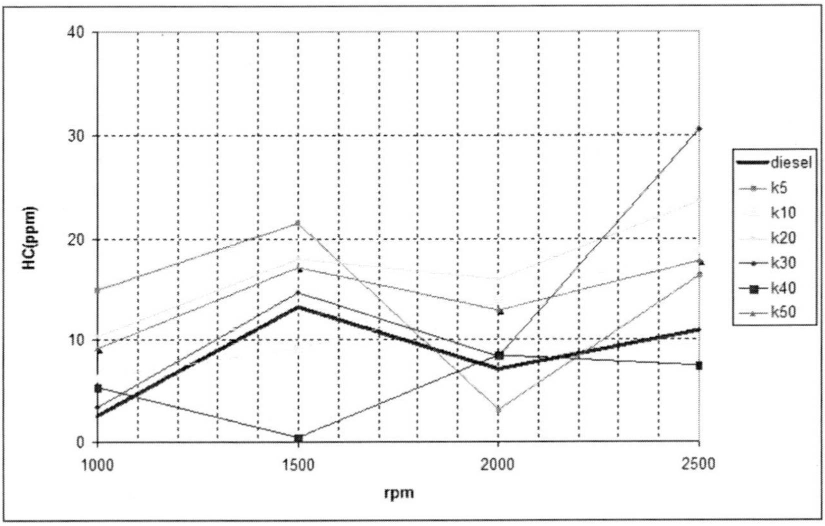

Figure 7. The HC variation on different rpm regarding to the mixture

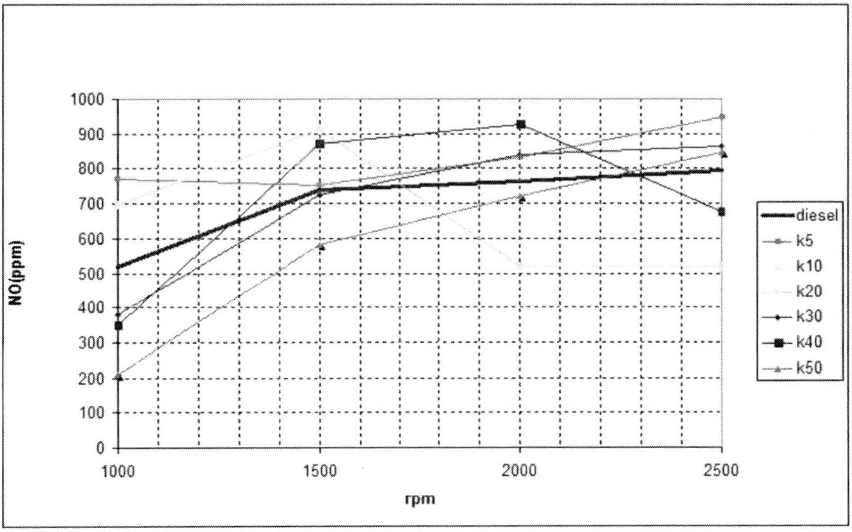

Figure 8. The NO variation on different rpm regarding to the mixture

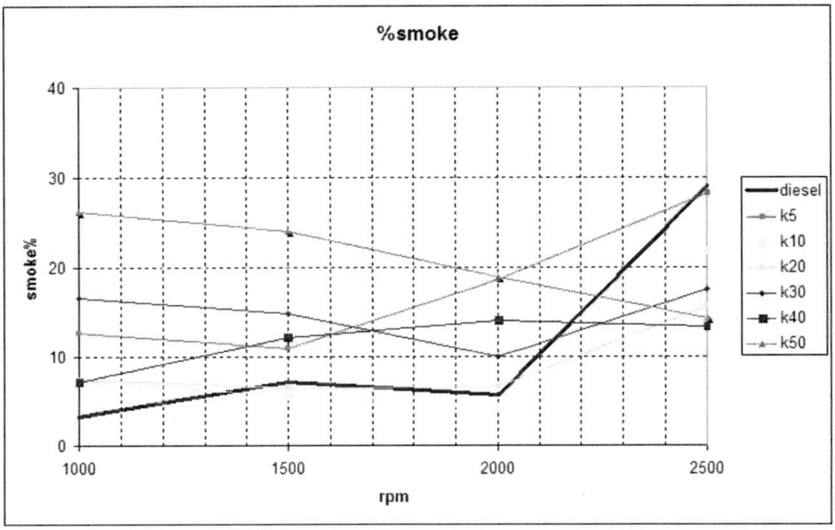

Figure 9. The smoke variation on different rpm regarding to the mixture

From figure 6 it is clear that when the maize oil is increased on the fuel regarding to diesel, it appears an increase of CO, except in the case k40/1500rpm. From figure 7 it can be noticed the biggest reduction of HC regarding to diesel in case of k40/1500rpm. From figure 8 it can be noticed the biggest reduction of NO regarding to diesel in the case of k20/2000-2500rpm. From figure 9 it can be noticed the biggest reduction for k10/1500-2000rpm. From the above figures it is clear that the use of different mixtures can constitute changes to CO, HC, NO and smoke too. It is also important the fact that there was no changes in the rounds of the engine, as well as in the supply of water at the use of mixtures. Finally as far as the consumption is concerned, did not observed changes with the use of different mixtures. The use of mixture of diesel and maize oil has the following impacts:

- About CO it can be noticed that when the maize oil is increased on the fuel regarding to diesel, it appears a decrease of CO, except in the case k40/1500rpm.
- About HC it can be noticed the biggest reduction of HC regarding to diesel in case of k40/1500rpm
- The biggest reduction of NO regarding to Diesel is noticed in the case of k20/2000-2500rpm.
- The smoke it can be noticed the biggest reduction for k10/1500-2000rpm

2.3. Cotton oil

In the experiment stage has been used directly cotton oil in the mixture of diesel in to a four – stroke Diesel engine and not elaborated in the figure of bio-diesel. Specifically it has been used diesel, mixture diesel- 10% cotton oil(B10), diesel- 20% cotton oil(B20), diesel- 30% cotton oil (B30), diesel-40% cotton oil (B40), diesel- 50% cotton oil (B50) in a four-stroke diesel engine [18]:

The experimental results are shown at the following tables and figures:

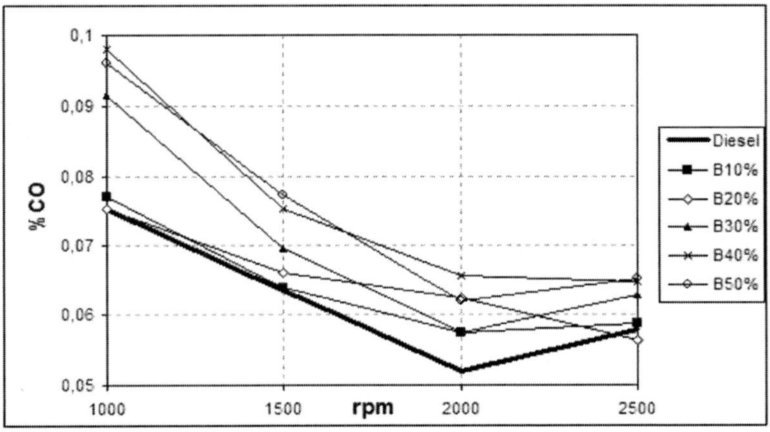

Figure 10. The CO variation on different rpm regarding to the mixture

From figure 10 it is clear that when the cotton oil is increased on the fuel regarding to Diesel, it appears an increasement of CO.

Table 9. The CO average value variation on different rpm regarding to the mixture

rpm	% CO					
	Diesel	B10	B20	B30	B40	B50
1000	0,075	0,076	0,075	0,091	0,098	0,095
1500	0,063	0,064	0,066	0,069	0,075	0,077
2000	0,052	0,057	0,062	0,057	0,065	0,061
2500	0,057	0,058	0,056	0,062	0,064	0,065

Table 10. The HC average value variation on different rpm regarding to the mixture

rpm	HC (ppm)					
	Diesel	B10	B20	B30	B40	B50
1000	30,78	35,86	39,04	39,05	14,86	46,64
1500	62,86	41,18	35,59	48,74	53,84	51,34
2000	125,52	83,84	101,38	109,07	76,42	142,94
2500	78,26	84,93	169,34	103,64	167,82	105,80

Table 11. The no average value variation on different rpm regarding to the mixture

rpm	NO (ppm)					
	Diesel	B10	B20	B30	B40	B50
1000	439,67	471,17	464,34	361,59	318,85	320,47
1500	649,65	660,83	626,78	611,71	565,26	522,16
2000	710,41	688,75	679,64	687,06	710,18	798,96
2500	868,88	930,50	919,53	919,08	987,35	947,80

Table 12. The %smoke average value variation on different rpm regarding to the mixture

rpm	%smoke					
	Diesel	B10	B20	B30	B40	B50
1000	7,72	5,76	6,36	13,89	12,88	13,35
1500	5,81	3,16	5,41	10,72	12,17	13,62
2000	5,24	3,62	4,45	7,59	7,28	7,70
2500	10,98	7,94	9,93	7,92	9,62	9,01

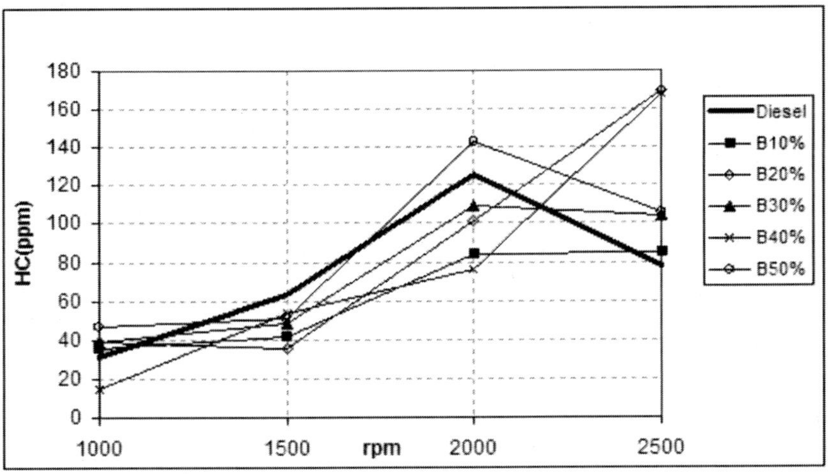

Figure 11. The HC variation on different rpm regarding to the mixture

From figure 11 it can be noticed the biggest reduction of HC regarding to Diesel in case of the mixture B20/1500 rpm and in the case of the mixture B40/2000 rpm.

From figure 12 it can be noticed the biggest reduction of NO regarding to Diesel in the cases of the mixture B40/1000 rpm, B50/1000 rpm and B50/1500 rpm too.

From figure 13 it can be seen the reduction of smoke regarding to Diesel in case of the mixture B10 and B20 at all rounds per minute. It can also be noticed the reduction of smoke in the case of B30, B40, B50/2500 rpm. Finally it can be seen an increasement of the mixture B30, B40, B50 at all rounds regarding to Diesel. From the above figures it is clear that the use of different mixtures can constitute changes to CO, HC, NO and smoke too.

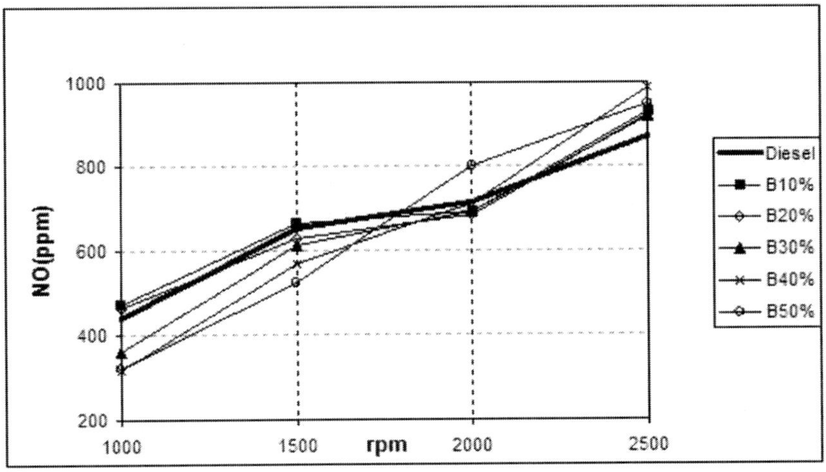

Figure 12. The NO variation on different rpm regarding to the mixture

It is also important the fact that there was no changes in the turns of engine, as well as in the supply of water at the use of mixtures. Finally as far as the consumption is concerned, did not exist changes with the use of different mixtures.The use of mixture of Diesel and Cotton Oil has the following impacts:

- About CO it can be noticed an increasement when the cotton oil is used as a fuel.
- About HC it can be noticed a reduction at 1500 rpm and particularly bigger reduction in the use of B20. It also appears reduction of the HC for all the mixture at 2000 rpm with the exception of B50. Finally about the HC, for all the mixture at 2500 rpm is observed increase of HC regarding to Diesel.

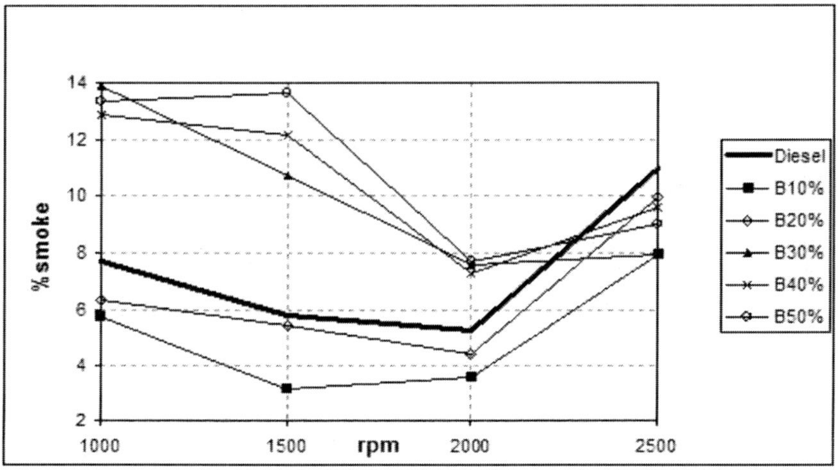

Figure 13. The smoke variation on different rpm regarding to the mixture

- About NO has been noticed a reduction at 1000 rpm and 1500 rpm for all the mixtures. A small reduction appeared for all the mixtures at 2500 rpm with the exception of B50, regarding to Diesel. Finally about the NO for all the mixtures appeared increase at 2500 rpm regarding to Diesel.

- About the smoke it can be noticed a reduction of the mixture of B20 and B10, but it appears an increasement for all other mixture in any round regarding to Diesel, with the exception of 2500 rpm, in where all the mixture appear a reduction.

2.4. Olive seed oil

In the experiment stage has been used directly cotton oil in the mixture of diesel in to a four – stroke Diesel engine. Specifically it has been used diesel, mixture diesel-5% olive seed oil (Pyrin5%), diesel-10% olive seed oil (Pyrin10%), diesel-20% olive seed oil (Pyrin20%), diesel-30% olive seed oil (Pyrin30%), diesel-40% olive seed oil (Pyrin40%), diesel-50% olive seed oil (Pyrin50%) in a four-stroke diesel engine [19]:

The experimental results are shown at the following tables and figures:

Table 13. The CO average value variation on different rpm regarding to the mixture

rpm	CO %						
	diesel	Pyrin 5%	Pyrin 10%	Pyrin 20%	Pyrin 30%	Pyrin 40%	Pyrin 50%
1000	0,056	0,056	0,054	0,060	0,053	0,053	0,048
1500	0,055	0,044	0,038	0,055	0,040	0,041	0,036
2000	0,043	0,038	0,031	0,050	0,031	0,036	0,030

Table 14. The HC average value variation on different rpm regarding to the mixture

rpm	HC (ppm)						
	diesel	Pyrin 5%	Pyrin 10%	Pyrin 20%	Pyrin 30%	Pyrin 40%	Pyrin 50%
1000	31,783	35,237	77,922	152,830	13,023	16,799	12,508
1500	38,001	48,434	79,198	165,479	22,954	24,870	22,860
2000	38,338	71,585	97,513	208,166	60,209	37,725	47

Table 15. The NO average value variation on different rpm regarding to the mixture

	NO (ppm)						
rpm	diesel	Pyrin 5%	Pyrin 10%	Pyrin 20%	Pyrin 30%	Pyrin 40%	Pyrin 50%
1000	518,210	415,212	375,075	392,478	372,681	473,620	362,663
1500	739,366	730,361	677,793	703,549	673,198	729,462	758,413
2000	762,155	790,676	738,929	805,702	825,376	938,210	880,990

Table 16. The %smoke average value variation on different rpm regarding to the mixture

rpm	%smoke						
	diesel	Pyrin 5%	Pyrin 10%	Pyrin 20%	Pyrin 30%	Pyrin 40%	Pyrin 50%
1000	9,990	12,605	14,787	12,717	11,018	9,932	16,278
1500	7,363	11,967	10,594	13,715	12,575	13,285	19,673
2000	6,634	14,212	12,201	14,131	14,098	17,528	23,359

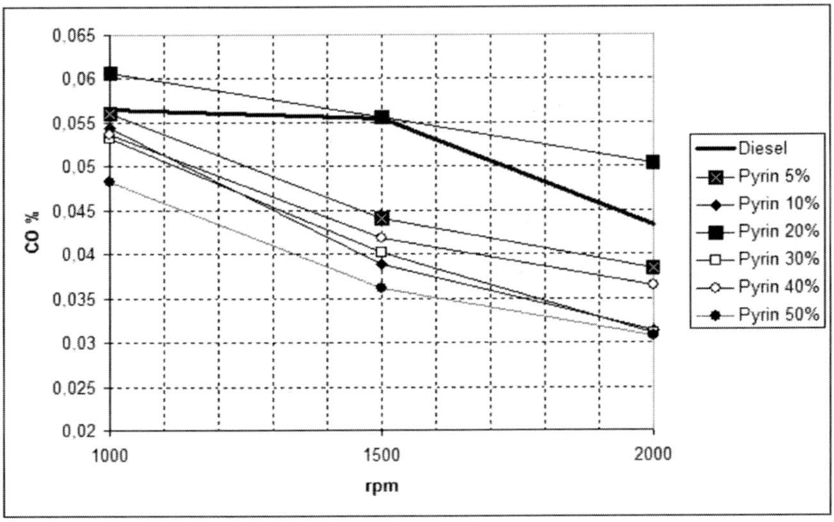

Figure 14. The CO variation on different rpm regarding to the mixture

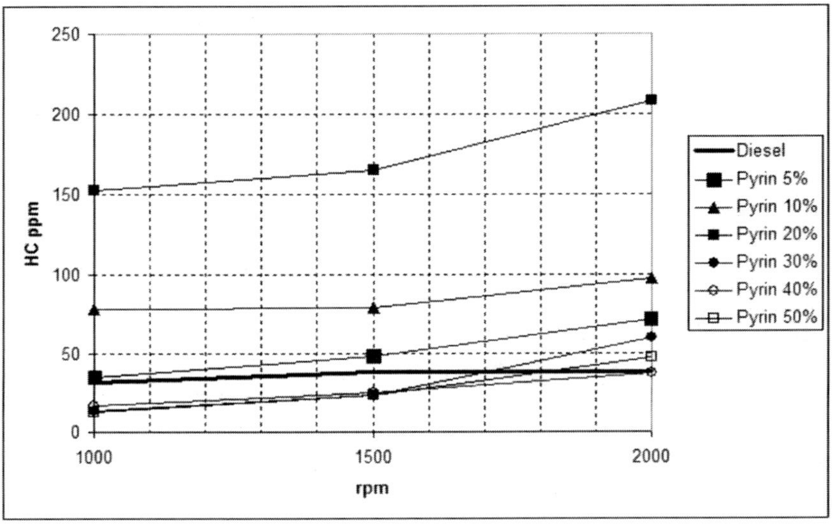

Figure 15. The HC variation on different rpm regarding to the mixture

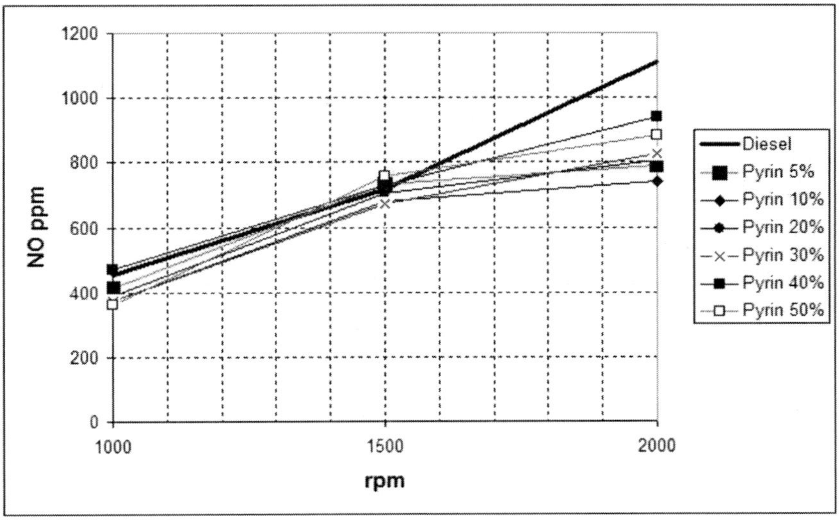

Figure 16. The NO variation on different rpm regarding to the mixture

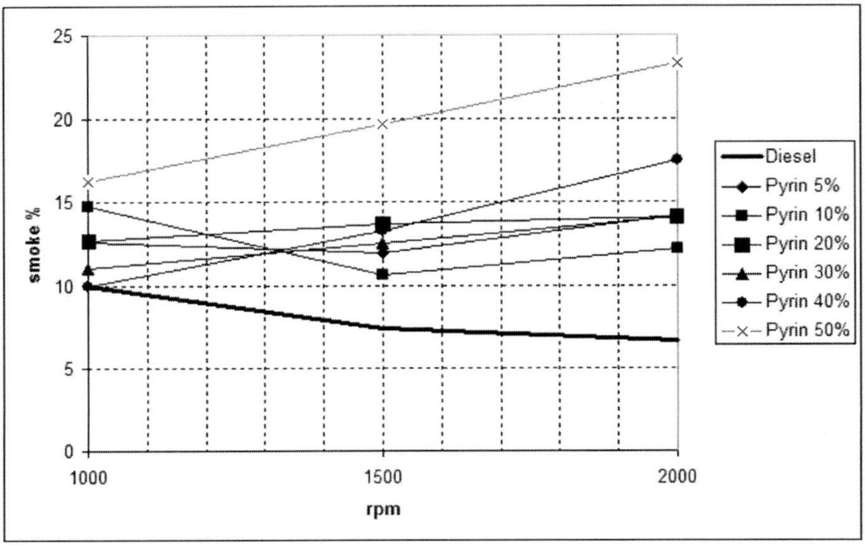

Figure 17. The smoke variation on different rpm regarding to the mixture

From figure 14 it is clear that when the olive seed oil is increased on the fuel regarding to diesel, it appears a decrease of CO. From figure 15 it can be noticed the biggest reduction of HC regarding to diesel in case of pyrin50%. From figure 16 it can be noticed the biggest reduction of NO regarding to diesel in the case of pyrin10%/2000rpm. From figure 17 it can be noticed that the best behaviour appears on diesel. From the above figures it is clear that the use of different mixtures can constitute changes to CO, HC, NO and smoke too. It is also important the fact that there was no changes in the rounds of the engine, as well as in the supply of water at the use of mixtures. Finally as far as the consumption is concerned, did not observed changes with the use of different mixtures. The use of mixture of diesel and olive seed oil has the following impacts:

- About CO it can be noticed when the olive seed oil is increased on the fuel regarding to diesel, it appears a decrease of CO
- About HC it can be noticed the biggest reduction of HC regarding to diesel in case of pyrin50%
- The biggest reduction of NO regarding to diesel in the case of pyrin10%/2000rpm.
- The smoke it can be noticed that the best behaviour appears on diesel.

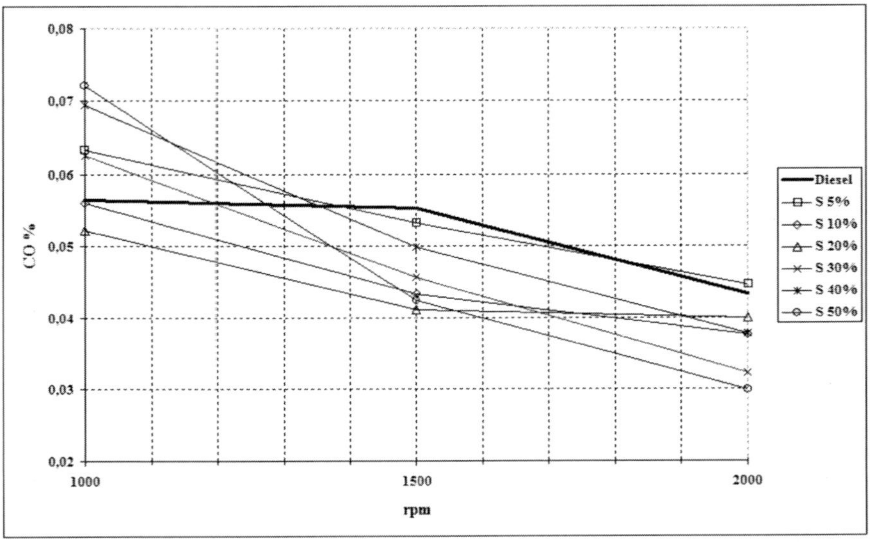

Figure 18. The CO variation on different rpm regarding to the mixture

2.5. Soy oil

In the experiment stage has been used directly soy oil in the mixture of diesel in to a four – stroke Diesel engine. Specifically it has been used Diesel, mixture Diesel-5% soy oil (S5), Diesel-10% soy oil (S10), Diesel-20% soy oil (S20), Diesel-30% soy oil (S30), Diesel-40% soy oil (S40), Diesel-50% soy oil (S50) in a four-stroke diesel engine [20]:

The experimental results are shown at the following tables and figures:

From figure 18 it is clear that when the soy oil is increased on the fuel regarding to diesel, it appears a decrease of CO, except in the cases S5,30,40,50/1000rpm.

Table 17. The CO average value variation on different rpm regarding to the mixture

rpm	HC (ppm)						
	Diesel	S5	S10	S20	S30	S40	S50
1000	31,78	21,15	21,88	8,28	5,76	54,61	28,01
1500	38,00	24,30	51,65	9,16	5,80	55,53	30,04
2000	38,33	23,70	89,90	28,68	22,34	84,88	67,47

Table 18. The HC average value variation on different rpm regarding to the mixture

rpm	NO (ppm)						
	Diesel	S5	S10	S20	S30	S40	S50
1000	454,2	387,6	397,5	416,1	414,8	341,0	277,9
1500	715,3	739,8	743,6	720,9	758,8	718,8	651,1
2000	1109,6	621,7	829,6	808,2	915,6	919,8	920,2

Table 19. The NO average value variation on different rpm regarding to the mixture

rpm	% smoke						
	Diesel	S5	S10	S20	S30	S40	S50
1000	9,99	8,72	9,41	11,61	14,26	18,32	24
1500	7,36	8,23	8,43	9,87	13,02	18,21	17,84
2000	6,63	6,25	7,70	8,08	11,27	17,21	20,5

Table 20. The %smoke average value variation on different rpm regarding to the mixture

rpm	CO %						
	Diesel	S5	S10	S20	S30	S40	S50
1000	0,056	0,063	0,056	0,052	0,062	0,069	0,072
1500	0,055	0,053	0,043	0,041	0,045	0,049	0,042
2000	0,043	0,044	0,037	0,04	0,032	0,037	0,029

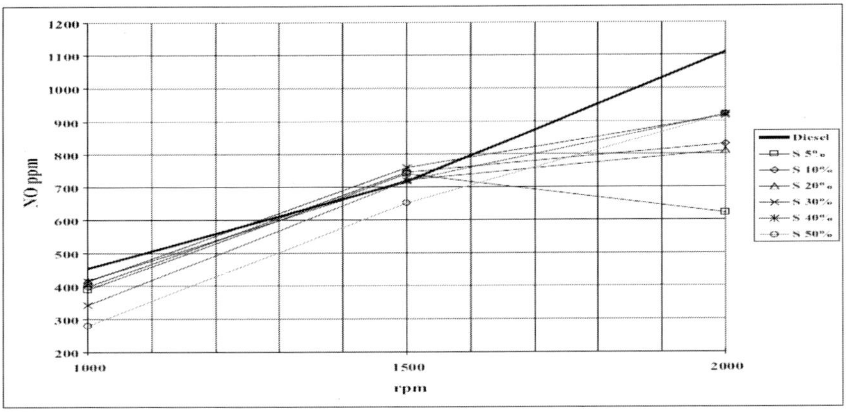

Figure 19. The HC variation on different rpm regarding to the mixture

From figure 19 it can be noticed the biggest reduction of HC regarding to diesel in case of the mixtures S5, S20 and the mixture S40.

From figure 20 it can be noticed the biggest reduction of NO regarding to Diesel in the case of the mixture S50.

From figure 21 it can be seen the increase of smoke regarding to diesel for all the mixtures. From the above figures it is clear that the use of different mixtures can constitute changes to CO, HC, NO and smoke too. It is also important the fact that there was no changes in the rounds of the engine, as well as in the supply of water at the use of mixtures. Finally as far as the consumption is concerned, did not observed changes with the use of different mixtures. The use of mixture of diesel and soy oil has the following impacts:

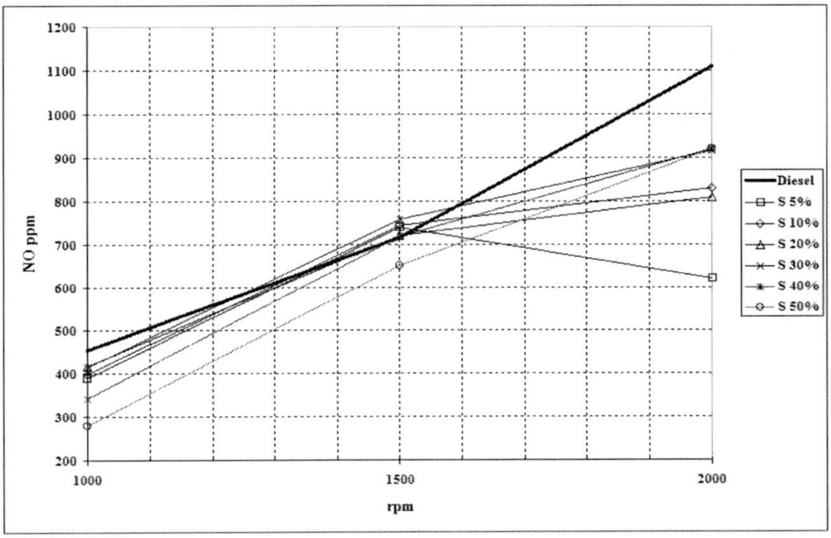

Figure 20. The NO variation on different rpm regarding to the mixture

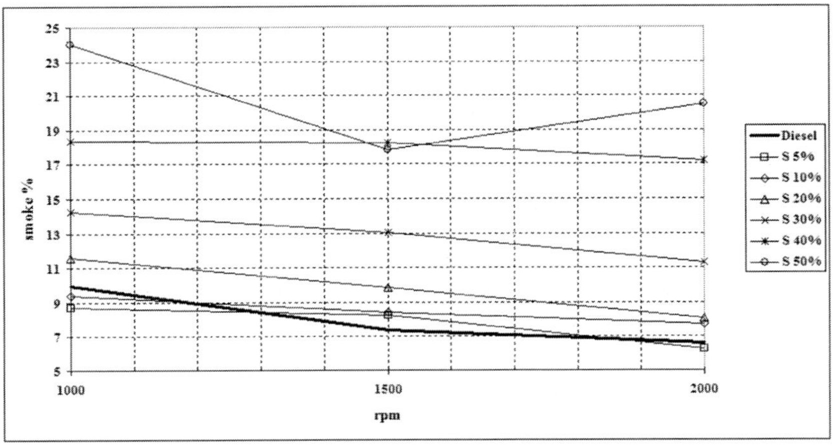

Figure 21. The smoke variation on different rpm regarding to the mixture

About CO it can be noticed that when the soy oil is increased on the fuel regarding to diesel, it appears a decrease of CO, except in the cases S5,30,40,50/1000rpm.

About HC it can be noticed the biggest reduction of HC regarding to diesel in case of the mixtures S5, S20 and the mixture S40.In the case of S30 appears the maximum increase of HC in relation to diesel.

The biggest reduction of NO regarding to Diesel is noticed in the case of the mixture S50.

The smoke is increased regarding to diesel for all the mixtures. Except the cases S5,50/1000rpm.

REFERENCES

1. Aldritton L, Monastersky R., Eddy A, Hall M, Shea E. (1992). Our Ozone Shield Reports to the Nation on Our Changing Planet. Fall 1992, University Cooperation for Atmospheric research office for interdisciplinary studies Boulder, Colorado
2. C. Arapatsakos, 2000Air and water influence of two stroke outboard engine using gasoline- ethanol mixtures, Transaction of SAE, Book SP-1565.
3. C. Arapatsakos, A. Karkanis, P. Sparis, Environmental Contribution of Gasoline- Ethanol Mixtures, WSEAS Transactions on Environment and Development, 7 2July 2006
4. C. Arapatsakos, A. Karkanis, P. Sparis, Gas emissions and engine behavior when gasoline-alcohols mixtures are used, Journal of Environmental Technology, 24 1069 1077
5. C. Arapatsakos, Testing the tractor engine using diesel- ethanol mixtures under full load conditions, International Journal of Heat & Technology, 19n.1, 2001
6. C. Arapatsakos, D. Christoforidis, A. Karkanis, The gas emissions variation of diesel engine from the combustion of used vegetable oils, Proceedings of International Conference on Energy and Environment, Cambridge 2009
7. Arapatsakos C., Christoforidis D, Karkanis A, Mitroulas D, Teka C. (2007). Test results from the use of cotton oil mixtures as fuel in a

four-stroke engine, International journal of Energy and Environment, Issue3 Vol. 1, 2007

8. Arapatsakos C, Christoforidis D, Karkanis A, Mitroulas D. Fuel of diesel- olive seed oil mixture′ International journal of energy Issue 3 vol 2, 2008.

9. C. Arapatsakos, D. Christoforidis, A. Karkanis, K. Mitroulas, Soy oil as fuel in a four stroke engine, Journal of wseas transactions on environment and development 2008

10. C. Arapatsakos, D. Christoforidis, G. Sarantitis, D. Giannopoulos, Fuel mixtures of diesel-maize oil International journal of energy 3 2 2008

11. F. Schafer, R. V. Basshuysen, 1995 Reduced Emissions and Fuel Consumption in Automobile Engines Published by SAE.

12. H. Menrad, M. Haselhorst, Alcohol fuels, Monograph. Springer, New York, ISBN 3211816968, 1981

13. Harrington A.; Shishu C. A Single-Cylinder Engine Study of the Effects of Fuel Type, Fuel Stoichiometry and Hydrogen-to-Carbon Ratio on CO, NO and HC Exhaust Emissions, SAE-Paper 730476

14. K. Owen, T. Coley, 1995Automotive Fuels Reference Book, (Second Edition), Published by SAE.

15. B. Mitchell, 1989The greenhouse effect and climate change. Reviews of Geophysics 27.

16. C. Arapatsakos, A. Karkanis, P. Sparis, Gas emissions and engine behaviour when gasoline-alcohol mixtures are used, Environmental technology 242003

17. Siddharth S., Green Energy-Anaerobic Digestion.Converting Waste to Electricity, WSEAS Transactions on Environment and Development, 7 2July 2006

18. Swedish Motor Fuel Technology Co., Alcohols and alcohol blends as motor fuels, Vol.II B, 8STU information 580 1986

19. T. Maxwell, C. Jones, 1995Alternative fuels: Emissions, Economics and Performance Published by SAE.

20. Schenewerk E, Automatic DRAC LMFBR to Speed Licensing and Mitigate CO2, WSEAS Transactions on Environment and Development, Issue 7, Volume 2, July 2006.

7

CHAPTER

SUSTAINABILITY ISSUES AND OPPORTUNITIES IN THE SUGAR AND SUGAR-BIOPRODUCT INDUSTRIES

Gillian Eggleston * and Isabel Lima

USDA-ARS-Southern Regional Research Center, 1100 Robert E. Lee Boulevard, New Orleans, LA 70124, USA

ABSTRACT

Like many other industries, the sugar and sugar-bioproduct industries are facing important sustainability issues and opportunities. The relatively low and fluctuating profit for sugar, surpluses of sugar, world-wide trend to produce alternative, renewable bio-based fuels and chemicals to those derived from petroleum and reduce greenhouse gases, water- and energy-intensive factories and refineries, and increased consumer demands for sustainably manufactured products are putting pressure on the industries to diversify for sustainability. Sugar crops, including sugar and energy cane (Saccharum officinarum), sugar and energy beets (Beta vulgaris), and sweet sorghum (Sorghum bicolor L. Moench), are excellent, renewable biomass feedstocks because of their availability, their being amongst the plants that give the highest yields of carbohydrates per hectare, and high sugar contents. While much research has been focused on conversion technologies for advanced biofuels and bioproducts, attention is now focused on developing sustainable supply chains of sugar feedstocks for the new, flexible biorefineries, with customers wanting maximum feedstock reliability and quality, while minimizing cost. All biomass from sugar crops are potential feedstocks. The cogeneration of bioelectricity from bagasse and leaf residues is being increasingly manufactured in more countries and, due to the high carbon content of bagasse and leaves, can also be converted into value-added products such as biochar. Sugar crops are superior feedstocks for the production of platform chemicals for the manufacture of a range of end-products, e.g., bioplastics, chemicals, and biomaterials. In several countries and regions, green sustainability criteria are now in place and have to be met to count against national biofuel targets. Processes to convert high-fiber sugar crop biomass into biofuel have been developed but there has only been limited commercialization at the large-scale.

Keywords: renewable sugar crops; biomass; advanced biofuels; bioproducts; bagasse; extraneous matter

1. INTRODUCTION

Sucrose (α-D-glucopyranosyl-(1 → 2)-β-D-fructofuranose) is ubiquitously known as common table sugar, and crystalline sucrose is primarily produced industrially from sugarcane (Saccharum officinarum) and sugar beet (Beta vulgaris) (Figure 1). Like many other food and chemical industries, the sugar industry and related sugar-bioproduct industries are currently facing tough sustainability issues. Sustainability is the balancing of the three, interdependent, development pillars of the environment (ecology), society, and economy (Figure 2). For some industries the core principles for sustainable manufacture are renew, reuse, and recycle, which are applied to every production step and business practice [1]. Sustainable development should also ensure that the needs of the present are met without compromising the ability of future generations to meet their own needs [2].

Figure 1. Sugarcane harvested into billets (**top**) and sugar beets being delivered for processing (**bottom**).

Although the twentieth century saw enormous growth in chemicals manufacturing which fed the parallel growth in the developed world, it came at a cost. Inefficient processes reliant on fossil fuels leading to unacceptable levels of pollution, hazardous operations resulting in a number of well-publicized disasters, inadequate product testing causing often irrational public concerns over product safety, have all led to an exponential growth in chemicals legislation [3]. Chemical industries, including the sugar and associated industries, are now working towards achieving environmentally acceptable and economically viable manufacturing in a tough legislative framework while meeting the high demands of a growing population. Sustainable production of sugar, biofuels (for example, first generation ethanol in Brazil) and other bioproducts (such as chemicals and structural materials) from sugar crops, will only be realized through a re-assessment of the entire chemical product life-cycle from resources, to manufacturing and production, through to product use and ultimate fate [3,4]. Moreover, several critical changes are required both in mindset and practice that are listed in Table 1.

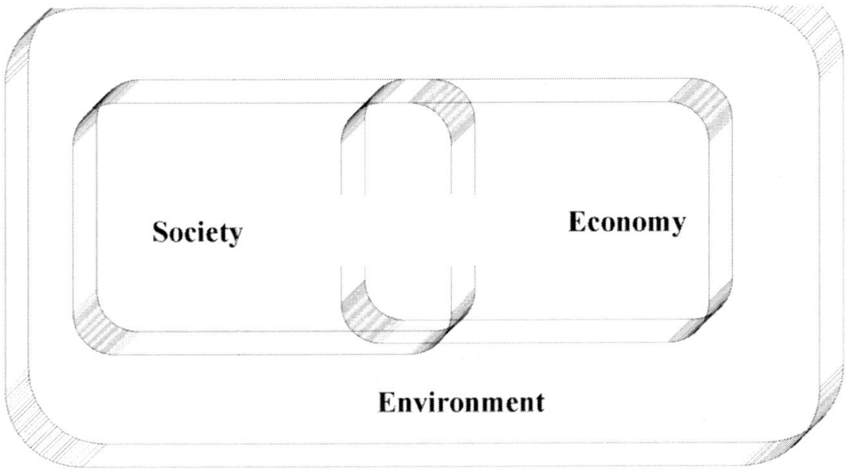

Figure 2. Sustainability focuses on the triple bottom line: (i) social responsibility; (ii) economic viability, which are both constrained by (iii) environmental limits that need protection. Sustainable development should recognize the well-being of human systems that is supported by a healthy, natural environment, and which future generations have an equal claim on our planets' resources.

Table 1. Unsustainable versus sustainable mindsets and practices in the current sugar industry. Adapted from [4].

Key Dimension	Unsustainable	Sustainable
Society/Policy Goals	Economic growth	Growth in well-being
Approach to Nature	Control over nature	Work with nature
Predominant Work Mode	Big is Better	Smart is Better
Focus on Business Activities	Goods	Services, needs
Energy Sources	Fossil fuels	Renewable energy (including biofuels and bioproducts)
Predominant Chemistry	Energy intensive	Low energy
Waste Production	High waste	No waste
Typical Materials	Iron, steel, and cement	Bio-based materials

Continued reliance on fossil fuel energy resources is unsustainable because of depleting world reserves and associated greenhouse gas (GHG), as well as energy security. This explains countries around the world legislating to curb GHG emissions. It also explains the currently vigorous initiatives of developing renewable and potentially carbon neutral, solid, liquid and gaseous biofuels as alternative energy resources as well as biobased alternatives to petroleum-derived chemicals and materials. Biomass, plant-derived organic matter, currently contributes to over 10% of primary energy to meet world annual demand [5] and is expected to grow further [6,7]. In the U.S., the Department of Energy (DOE), via the U.S. Biomass Roadmap, put forward the goal that by 2030 biomass will supply energy approximately equivalent to 30% of current petroleum consumption [8]. The European Commission (EC) has set mandatory targets for an overall share of 20% renewable energy and a 10% share of renewable energy in transport by 2020 [9]. The EC also reached an agreement in 2014 on an indirect land-use change (ILUC) directive [10], to minimize the impact of indirect changes of land use (for biofuel use), while at the same time protecting existing investments in biofuel production in Europe.

Two major challenges that need to be overcome for the creation of renewable biofuels and bioproducts from sugar crops are the need for: (i) net energy gain during feedstock production; and (ii) GHG emissions that are lower than those from fossil fuels [7]. This paper describes current trends, needs, and opportunities in the sugar and sugar-bioproduct industries that are expected to strongly contribute to their sustainability.

2. BACKGROUND INFORMATION: INDUSTRIAL PRODUCTION OF SUGAR AND ASSOCIATED BY-PRODUCTS

2.1. Crystalline Sugar Manufacture

Commercially available sucrose has very high purity (>99.9%) making it one of the purest organic substances produced on an industrial scale. To obtain such a pure product from both sugarcane and sugar beet, complex isolation and purification process units are followed. Industrial sucrose production is essentially a series of separations of non-sucrose compounds (impurities) from sucrose (Figure 3). Sugarcane is grown in tropical and sub-tropical areas of the world and processing often occurs in two stages. Juice is first extracted from sugarcane (sucrose yields range between 10%–15% weight of sugarcane) by tandem milling or diffusion and converted to raw sugar (~97.5%–99.5% pure sucrose; golden yellow/brown crystals) at factories. Secondly, after raw sugar has been transported to a refinery, it is refined using similar unit processes used in raw sugar manufacture, as well as additional decolorization steps such as ion-exchange resins and activated carbon, to the familiar white, refined sugar (>99.9% sucrose) [11].

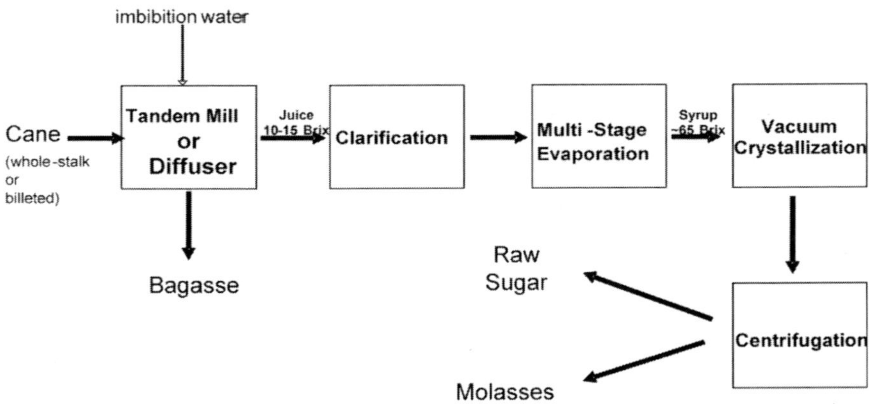

Figure 3. Basic scheme of the raw sugar manufacturing process in a sugarcane factory [11]. Brix is % dissolved refractometric solids.

Sugar beets are grown in more temperate areas and processed directly into white sugar (>99.9% sucrose) at nearby factories. Production of refined sugar from sugar beets has some similarities to refined cane sugar production, but dissimilarities exist because sugar beet is a tuberous root and sugarcane a grass. Sugar beets are introduced to the factory, washed, and sliced into "V" shaped cossettes. Sucrose and impurities are extracted from cossettes with hot water in a diffuser. Diffusion juice contains ~12% sucrose and 2% soluble impurities on sugar beet weight, and is heated to ~85 °C before it is clarified. The resulting "thin" juice is then concentrated from ~14 to 60–65 Brix (% dissolved refractometric solids) syrup or "thick juice" across multiple-effect evaporators, then triple-crystallized and centrifuged to produce white, refined sugar (>99.7% purity). As in sugarcane refineries, some sugar beet factories employ additional purification steps. For more detailed information on the industrial production of sucrose from sugarcane and sugar beet, the reader is referred to other comprehensive texts [11,12,13,14,15].

2.2. By-Products of Sugar Manufacture

The major by-products of crystalline sucrose manufacture are sugarcane bagasse, beet pulp, and sugarcane/beet molasses. The major agricultural residue is sugarcane extraneous leafy material. Minor by-products include fly ash, filter cake, lime and calcium carbonate residues. By volume, fibrous bagasse is the most important by-product and is the primary source of fuel for the generation of steam and electricity to operate sugarcane factories (see Section 4.1). Wet and dry beet pulp as well as pressed pulp silage, with or without added molasses, are sources of animal feed. Molasses is a valuable by-product of sugar manufacture and exists in a range of grades: edible molasses, cane and beet molasses, and refinery molasses. It is used as an animal feed additive, in the industrial production of rum and other beverage alcohols, bakers' yeast, citric acid, and other fermentation processes [4].

3. SUSTAINABLE SUPPLY CHAINS OF SUGAR BIOMASS FEEDSTOCKS FOR THE MANUFACTURE OF ADVANCED BIOFUELS AND BIOPRODUCTS

3.1. Sugar Crops as Biomass Feedstocks

A major trend in the U.S. and world-wide is to manufacture advanced biofuels and bioproducts from sugar crops, including sweet sorghum (Sorghum bicolor L. Moench), energy cane (Saccharum officinarum) and energy beets (Beta vulgaris). Sugar crops and their associated by-products and residues make good renewable carbohydrate feedstocks because they are readily available, can be grown in a much larger area of the world than grain crops [4], and are amongst the plants giving the highest yields of carbohydrates per hectare. A unique advantage sugar crops have over grain and cellulosic crops, are that they require less processing as their juice sugars are directly fermentable. Moreover, concerns of conversion efficiency, high production costs, and increased prices for whole corn grain and dried distiller's grain in the USA have provided a real opportunity for alternative feedstock sources from sugar crops [16]. Sugarcane and sugar beet industries, furthermore, have well-established agricultural production systems with a well-developed logistics and processing structure.

At present, approximately 40% of the world's fuel ethanol production is already from sugar crops (mostly the fermentation of either sugarcane juice or molasses in Brazil, Thailand, and India) with the remaining 60% from grain crops [17]. Thus, as of 2014 most fuel ethanol manufactured around the world is first generation ethanol. Also, since 2011, world fuel ethanol production has been steadily rising, with the latest reported figures for current production of 24,570 million gallons in 2014 [18]. The Environmental Protection Agency of the U.S. government designated Brazilian sugarcane ethanol as an advanced biofuel in 2010 due to its estimation that 61% reduction of life cycle GHG emissions, including direct and indirect land-use change emissions [19]. It has been acknowledged, however, that first-generation fuels have technical issues, such as their oxygen contents, that limit their use because they cannot completely replace fossil fuels on their own. For example, 2-carbon ethanol has lower energy per volume than petroleum fuels, and is not fully compatible with existing vehicles and the current fuel-distribution infrastructure [20]. Biofuels with high carbon contents, e.g., 15-carbon alkene β-farnesene that is converted from sugar by engineered yeasts by Amyris in California, offer more sustainable substitution of fossil fuels in the future. Moreover, β-farnesene addresses reduction in GHGs while also delivering improved engine performance [20]. Very recently, Balakrishnan et al. [21] went even further and showed that sugars from sugarcane can be converted in >95% yields to a new class of cycloalkane compounds used for aviation fuels and also achieved net life cycle GHG savings of up to 80%.

While much research has been focused on conversion technologies for advanced biofuels and bioproducts, attention is now focused on developing sustainable supply chains of sugar feedstocks for the new, flexible biorefineries [22]. This includes improved feedstock quality and cost-effective approaches for minimizing feedstock sugar losses during storage [22]. Kenney et al. [23] similarly reported that while "much progress has been made in improving biomass collection and pre-processing machinery performance and efficiencies, reducing material losses throughout the supply chain, and expanding harvesting and storage operational windows" and "emphasis on feedstock quality is still lacking." The quality of the feedstock supply chain from the field to the processed end-product is impacted by crop genotype variability, production conditions, harvest method, collection and storage practices, season date, and environmental conditions. These relationships and their integration are illustrated in Figure 4 using the examples of sweet sorghum and energy beets, but other sugar crops such as energy cane could be included as well as by-products and residues. Furthermore, as the current pioneer biorefineries move from technology development and deployment to operations, the quality and specifications of the feedstock will become even more important [23]. Overall, customers of sugar feedstocks want maximum feedstock reliability and quality, while minimizing variability and cost.

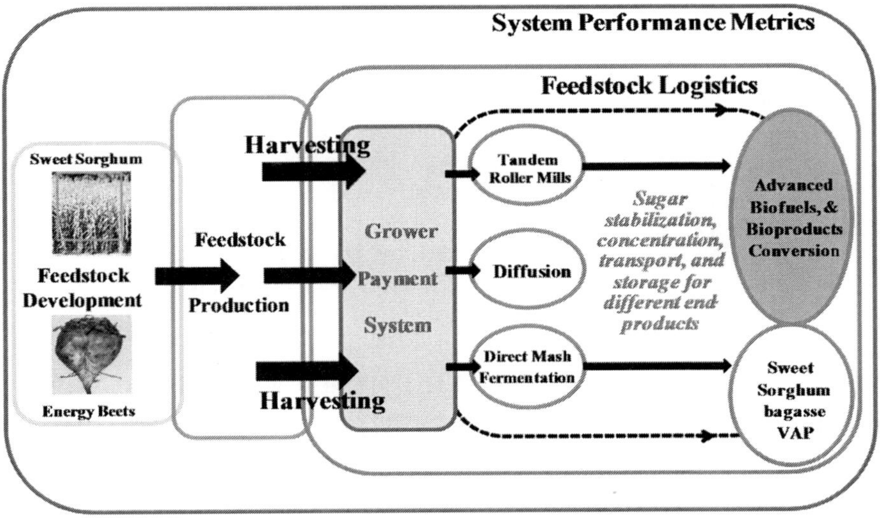

Figure 4. The manufacture of advanced biofuels and bio-products supply chain components for sweet sorghum and energy beets. VAP = value added products. This schematic is also applicable to energy canes, bagasse, and extraneous leafy matter.

3.2. Sweet Sorghum as a Biomass Feedstock

Sweet sorghum is a type of sorghum that, like sugarcane, contains juice rich in soluble sugars as well as fiber but, unlike sugarcane, also contains starchy grain that could be used for food, animal feed, or non-food products. Sweet sorghum is an attractive biomass feedstock because it can overcome many of the shortcomings of other sugar crops, due to its efficient C4 photosynthetic pathway, easy cultivation from seed, low fertilizer and water requirements, growth on marginal lands, short growth cycle (90 to 150 days after planting, depending on cultivar and environment) that allows the possibility of multiple crops per season or rotation with other food or non-food crops, wide geographical suitability, and huge breeding potential [24]. It must be noted, however, that due to its low juice purity (sucrose/Brix × 100) of only ~75%, which is markedly lower than for sugarcane (~85%) and sugar beets (~87%) as well as high reducing sugar content, it cannot be used to manufacture crystalline sucrose [25]. Although many current global applications of sweet sorghum are still for small-scale edible syrup and forage production, there has been a recent dramatic increase in interest for large-scale biofuel and bioproduct manufacture [24,26,27]. Several private-sector groups in the U.S. and world-wide have been pursuing development of new domestic industrial sugar feedstocks from sweet sorghum to supply the bioprocessing demand, and new biorefineries are now becoming a reality. For example, Heckemeyer Mill in rural Sikeston, Missouri, recently built the largest, operational sweet sorghum biorefinery in the U.S., which is capable of crushing up to 82 tonnes/h (equivalent to ~49 ha/day) and producing 24,000 gal of syrup (80% solids)/day [28]. Moreover, as sugarcane factories can sit idle for up to 9 months of the year, processing of sweet sorghum to syrup in factories before the sugarcane harvest would allow the use of under-utilized capital equipment. This is currently being explored and tested in numerous sugarcane factories/distilleries in Brazil and by Okeelanta, Florida [29].

4. BAGASSE FROM SUGAR CROPS: LARGE-SCALE COGENERATION OF ELECTRICITY AND OTHER ENERGY PRODUCTS

4.1. Cogeneration of Electricity from Bagasse

Bagasse, the fibrous fraction remaining after juice extraction from sugarcane and sweet sorghum, varies in composition depending on intrinsic and extrinsic factors such as genotype, maturity (lignifications), and environment [30]. Sweet sorghum bagasse is available to a much lesser extent than sugarcane, and its higher amounts of protein can make it more valuable as an animal feed [24]. Most sugarcane processors burn bagasse to cogenerate steam and electricity for running the factory [31]. The recovery of energy from bagasse is also a major reason that sugarcane has a higher net energy ratio (output/input) than many other crops [7]. Furthermore, cogeneration contributes to sustainability as the negative environmental impact of GHGs from traditional thermal power stations are reduced [32]. Presently, some countries' sugar industries, e.g., Brazil, Mauritius, India, Australia, El Salvador, Nicaragua, Guatemala, Columbia, and The Philippines, also operate large-scale cogeneration of electricity and sell the surplus to the local or national grid, and there is great potential for many other countries to follow. Currently, bagasse-based cogeneration is the most profitable sector in the large Brazilian sugar industry, with underperforming sugar and ethanol sectors being bailed out and/or supported recently by cogeneration units [33].

The success of the cogeneration of bagasse depends on the availability of adequate technology as well as a profitable price structure [34]. In recent years, there have been major technological improvements leading to higher efficiency cogeneration of electricity from bagasse, in particular the use of new high-pressure boilers, i.e., up to 82–100 bar (producing superheated steam at 525 °C) [35]. Efficiency gains leading to a surplus of electricity generation for export to the grid have also been accomplished through the retro-fitting of turbo-alternators with high steam pressure and temperature [31], and the optimization of other process parameters, including steam consumption, increasing fiber content of sugarcane through breeding, lower moisture content of bagasse, and reducing the consumption of electricity in the factory tandem mill and power plant [32]. In Brazil, during the 2009/2010 harvesting season the total electricity produced from sugarcane bagasse was 20,031 GWh and this value may rise up to 68,730 GWh over the next 9 years, as long as all factories install 99 bar boilers and 1.04 billion tons of sugarcane is produced [35].

4.2. Second Generation Biofuels from Bagasse

The sustainable trend towards electricity generation from bagasse has been accompanied by progress in developing other large-scale uses for the material, e.g., second-generation ethanol and other biofuels [30] and syngas from Fischer-Tropsch (FT) processes [36,37]. Lignocellulosic ethanol involves the conversion of cellulose and hemicelluloses, which is much more difficult than the conversion of starch and simple sugars. On the other hand, ethanol produced from lignocellulosic feedstocks is seen as a viable option to decrease any perceived competition between the production of foods and bioenergy [30], although energy will always be needed for food production. Other advantages over first generation ethanol are: (i) lignocellulose and cellulose are abundant and less expensive than agricultural food feedstocks; (ii) growth potential is huge; and (iii) some cellulosic crops can be grown in marginal lands that often require less fertilizer and water inputs but grow better on good quality land with optimal imputs.

There are five key-steps for the production of lignocellulosic ethanol from bagasse: (i) milling; (ii) chemical or physical pre-treatment; (iii) enzymatic hydrolysis or saccharification; (iv) fermentation of hexose (C6) and pentose (C5) sugars; and (v) distillation-dehydration of the ethanol. However, the processing technology for conversion in the most part has only reached very limited commercial scales. Dedini S/A Indústrias de Base, was the first company to build a bagasse-based pilot-scale ethanol facility in Brazil, but commercialization has yet to take place [38]. Gran Bio established the first second generation ethanol facility in Brazil in 2014 in Alagoas. Raizen now has a second generation ethanol facility in Piracicaba, Brazil, that started operating from July 2105. The commercialization of second generation bioethanol depends mostly on economic factors such as values for agricultural feedstocks that have been estimated to range between 50%–80% of the total ethanol's cost [39], government tax incentives for ethanol production, and mandatory ethanol/gas blends [40]. Dantas et al. [41] in their assessment of the productivity and cost of different technological routes, determined that, when compared to ethanol production, burning bagasse to generate electricity provided the most benefits from an investor perspective, although this may not be applicable to every country. For more information on this lengthy topic see [41].

4.3. Biochar

Biochars (Figure 5) can be produced via the thermo-chemical conversion of organic feedstocks. Thermo-chemical conversion reactions encompass processes such as: (1) gasification; (2) slow pyrolysis; (3) fast pyrolysis; (4) hydrocarbonization; and (5) combustion, and distinctions among these different options are related to the relative availability of oxygen, residence time, temperature and pressure. Slow pyrolysis (residence time is minutes or hours), in particular, is feedstock-flexible, and this has resulted in the research of a multitude of biomass materials as possible precursors due to their lower cost and availability. Yaman [42] published a review of pyrolysis of various biomass types which includes pyrolysis conditions. Sugarcane bagasse, due to its high carbon content, can serve as an excellent biochar feedstock. Within the thermo-chemical platform, pyrolysis generates biochar as the main product, synthesis gas and bio-oil from the non-condensable fraction of the gas with the split between liquid, char and gas being governed to a certain degree by variation in process conditions. During thermal decomposition of the organic material under limited supply of oxygen and a regime of high temperature (300–700 °C), the material undergoes a series of cleavage reactions, as volatile matter evolves, resulting in a porous high carbon product (Figure 6), mainly composed of aromatic compounds characterized by rings of 6-C atoms linked together. Biochar is part of the black carbon continuum with variable properties due to the net result of production (e.g., feedstock and pyrolysis conditions) and post-production factors (storage or activation). Therefore, biochars are not a single entity but rather span a wide range of black carbon forms [43].

Figure 5. Biochar in various forms (pelletized, granular [18 × 40 mesh] and powdered [less than 100 mesh]).

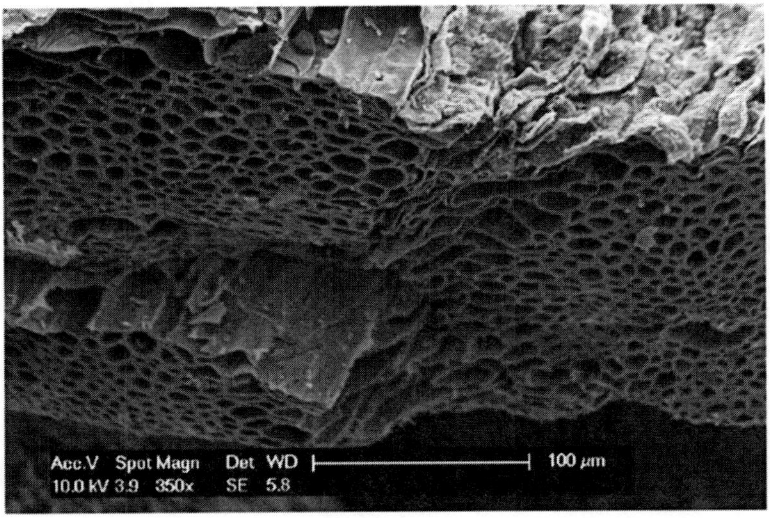

Figure 6. Scanning Electron Micrograph of a sugarcane bagasse biochar (magnification of 350×), showing its highly porous nature.

The terminology of biochar is commonly associated to its use in soil applications, and is unlike charcoal, which traditionally was manufactured from coal, and used as a fuel. The first use of the term biochar was around 1998 for the solid residual of biomass pyrolysis [44]. Biochar can be further upgraded or converted into activated carbon through either physical or chemical activation to produce a highly porous material that can be used in adsorption applications.

4.3.1. Biochar as a Soil Amendment

Biochar has been applied to soils virtually from the dawn of civilization, since fire pits were built on soil [43]. The current application of biochar to soil has been modeled after the Amazonian Terra Preta soils, which have higher soil fertility believed to result from intentional additions of biochar from "slash and char" agricultural practices [45,46,47]. Resulting biochars' nutrient contents are variable because they are based on feedstock and production conditions, however, differences have been noted in the chemistries of various biochars despite the similarity in production conditions [48]. Biochar consists mostly of carbon, hydrogen, and oxygen, but also contains the majority of the inorganic micronutrients that were native to the original feedstock. Thus, for biochars made from sugarcane bagasse, the mineral content is transferred to the biochar and concentrated due to burn-off of carbon moieties.

Inorganic material located at the surface of the biochar contributes to surface functional groups and charge that can advantageously enhance microbial relationships in soil as well as contribute to soil's nutrition [49]. Microbial biomass appears to increase in the presence of biochars, however, there is still little insight into the mechanisms by which biochar influences soil microorganisms, fauna, and plant roots [50]. When applied to soil, biochars have also been shown to suppress GHGs such as methane and nitrous oxide from soil [51]. In a recent greenhouse study, Lima et al. [52] observed that, by adding 4% v/v sugarcane bagasse biochar to soil, the sugar yield of sugarcane improved by 25% over the control. Furthermore, significant increases in sugarcane crop yield were also observed.

4.3.2. Biochar as Fuel

Pyrolysis of biomass including both agricultural and processing residues, as well as forestry products, has been traditionally used to make fuel or charcoal. When considering fuel production, densification technologies such as pelletization and agglomeration can improve the properties of the biomass feedstocks while generating a harder, less abrasive, and higher quality denser fuel. Zandersons et al. [53] produced charred fuel briquettes from sugarcane bagasse and determined that volatile matter given off during pyrolysis was sufficient to provide drying heat as well as sustain the pyrolysis process. Molasses is commonly used as a binder and conveniently available at the factory, making co-location of biochar manufacturing facilities in proximity to the factory even more efficient. Because sugarcane bagasse is considered a softer biomass material when compared to other biomass materials such as wood, Pendyal et al. [54] used sugarcane and corn syrup molasses as binders to create briquettes before pyrolysis. The suitability of bagasse biochar as a fuel, either by itself or blended with other by-products (e.g., leafy material) depends, among other parameters, on its heating value. A high carbon content, reported as 47% on a moisture and ash free basis, is also desirable and directly affects the fuel value [55], while ash content negatively impacts fuel value. Several empirical correlations have been established to calculate heating value from either proximate analysis (volatile matter, fixed carbon and ash content) [56] or from the elemental composition (e.g., C, H, N, O) [57]. Based on several available reports in literature, dry bagasse contains 82%–87% volatile matter, 1%–7% ash, and 10%–17% fixed carbon [58,59,60,61].

4.3.3. Biochar as an Adsorbent

Extensive literature exists on the use of biochars and their activated counterparts in remediation applications. Biochars produced from biomass such as sugarcane bagasse have low surface area when compared to commercial activated carbons, but their surface functionality can make them an attractive alternative particularly in adsorption of charged species. Additionally, they have a relatively lower cost of production, due to higher yield and more simple manufacture. When studying various agricultural by-products blended with various types of molasses, Ahmedna et al. [62] found that sugarcane bagasse was a better choice than rice straw or rice hulls as a precursor for activated carbons with desirable sugar decolorization properties, especially when combined with corn syrup. In another study, Ahmedna et al. [63] found that bagasse activated carbons performed as well as commercial carbons commonly used in sugarcane refineries, despite the fact of having half as much surface area. Because color removal both at the sugarcane factory and particularly refinery is such an integral part of the process, it is logical from a sustainability viewpoint to utilize biochars from sugarcane bagasse and/or trash as alternate color removal adsorbents. Feasibility studies for producing activated carbons from sugarcane bagasse and molasses revealed the cost to be $3.12/kg of carbon for a 2000 kg daily output [64,65]. In heavy metal ion adsorption applications, a sugarcane bagasse activated carbon outperformed a commercial carbon in the adsorption of nickel, copper, zinc, and lead, despite having much lower surface area [66]. This was most likely because of the presence of surface oxides such as carbonyls, lactones, phenols, and carbonyls giving biochars a negative surface charge [63]. Binder choice can also play a significant role in the biochar properties although properties of the base material will ultimately determine the properties of the biochar and the activated carbon [54].

4.3.4. Biochar for Carbon Sequestration

The process of biochar production has the potential of being carbon negative and, therefore, a carbon sinking technology by which a large percentage of the carbon in the biochars is in the fixed form. As sugarcane grows, it takes in carbon dioxide and converts it into plant building blocks such as cellulose and lignin, which will be again turned back into carbon dioxide through combustion. This process can be altered if carbon in sugarcane bagasse is instead converted into fixed carbon form as happens with pyrolysis. Pyrolysis temperature and residence time are directly proportional to the percent of fixed carbon generated. To effectively reverse climate change it may be necessary to return some of the atmosphere's carbon dioxide to the soil [67]. One way of achieving this, is by pyrolyzing biomass such as sugarcane bagasse to produce gas or oil for energy and using the remaining chars as soil amendments. Benefits of biochars use as climate change mitigation tool was recognized and reported to the U.S. Congress [68].

4.4. Storage of Large Piles of Bagasse

Challenges still exist with respect to the prolonged storage of very large quantities (can be >900,000 tonnes) of bagasse (~50% moisture content), particularly for off-season electricity generation. These include microbial decomposition and loss of fuel value, self-heating leading to spontaneous combustion, bagasse handling, and a variety of health issues, and environmental impacts [37]. An important health issue is bagassosis—A respiratory disease resulting from exposure to fungal spores from moldy bagasse dust. Environmental impacts include: (i) water pollution due to run-off from bagasse piles; (ii) bagasse dust that is a major problem in windy areas and at bagasse transfer points; (iii) noise and lights that have caused problems at storage sites close to residential areas; and (iv) off-odors.

Studies have been undertaken in various countries to overcome storage challenges, and in the case of Australia, guidelines for storage have even been established with cooperation of the Environment Protection Agency [37]. Options to solve bagasse storage problems have been reported and include the patented "Bagatex-20" process developed in Brazil [69], which depends on a biochemical catalyst to accelerate and control fermentation in the piles. Other options are forced drying before storage [70], special stacking, covering of piles [37], and densification processes. The latter includes pelletizing, which is known to significantly reduce final volume with bulk densities in the range of 1030 to 1260 kg/m^3, depending on pellet diameter [55]. Furthermore, pelletizing sugarcane bagasse is a way of improving fuel handling, transportation, conversion and also allowing for storage for off-season utilization [55]. Recent initial results from a sugarcane bagasse storage study in the U.S. [71], indicated that dramatic gains in fuel value can be achieved just by covering stored bagasse piles with tarpaulin (Table 2). Covering the bagasse also reduced the rate of deterioration [71].

Table 2. Proximate Analysis and estimated fuel value of sugarcane bagasse samples as a function of storage conditions across two factories *.

Storage Conditions	Sample	Moisture	Ash Content	Fixed Carbon	VOC	Fuel Value, KJ/kg	Moisture
		%	%	%	%	HHV [59]	LHV [72]
Covered	Fresh	38.5	6.0	15.7	78.3	17,712	10,890
	3 months	49.9	8.3	13.6	77.7	16,980	8504
	4 months	43.1	9.0	13.9	77.1	16,857	9590
	5 months	33.6	7.6	15.3	77.1	17,371	11,535
	6 months	56.9	9.9	14.0	76.1	16,738	7213
Uncovered	Fresh	-	-	-	-	-	-
	3 months	81.0	15.7	8.9	75.5	14,779	2805
	4 months	73.7	10.9	11.2	77.9	16,021	4203
	5 months	73.1	17.5	8.8	73.7	14,472	3889
	6 months	68.1	15.2	11.1	73.8	15,289	4875

* Data presented at the 45th Annual Joint ASSCT Meeting in New Orleans, LA 2015 [71].

5. NEW SUSTAINABLE USES FOR SUGAR CROP EXTRANEOUS MATTER (LEAVES AND TOPS)—AGRICULTURAL BIOMASS RESIDUE

5.1. New Uses for Sugarcane Extraneous Matter

Although sugarcane extraneous matter (E.M. also known as trash or leaves and tops) is an agricultural biomass residue that is widely available in many countries, but is still under-used or improperly utilized. There is, however, a growing reality that E.M. represents a rich source of renewable biomass. Sweet sorghum E.M. is also available but to a much lesser extent than for sugarcane, and the higher amounts of protein in sweet sorghum can make it more valuable as an animal feed [24]. For sugarcane, the continuing global reduction in pre-harvest burning of cane leaves as well as the increased use of green mechanical harvesting have considerably increased the availability of E.M. that can be collected either in the field or at the factory [35]. It should be noted, however, that some sugarcane E.M. must be left on the fields to protect the soil and recycle nutrients [73,74], nevertheless there is still plenty left to serve as a biomass feedstock.

Like other sustainable biomass sources, E.M. can be converted through a wide range of biochemical or thermal platform technologies into a multitude of chemicals, bioenergy, and biomaterials including cellulosic ethanol, electricity, and biochar [75]. E.M. also represents a real opportunity for rural communities to access local energy supplies and bring economic opportunities to many developing countries that grow sugarcane [5]. Sugarcane E.M. has already contributed to greater cogeneration of electricity [31,35]. In 2007 a study in Brazil showed that the utilization of sugarcane E.M. with bagasse can double the MWh production of electricity compared to bagasse alone [76]. E.M. can also be combined with bagasse to produce biochar, either with or without pelletizing, and the presence of certain minerals in the E.M. such as calcium, potassium, phosphorous, can be beneficial when using the biochar as soil amendment in sugarcane.

Considerable amounts of E.M. are still delivered to factories and processed to the detriment of the quantity and quality of raw sugar produced [77]. Thus, the development of large-scale industries utilizing E.M. as a biomass feedstock could additionally improve the sustainability and profitability of the sugar industry by indirectly improving the quality of raw sugar. The use of sugarcane E.M. as a biomass feedstock is greatly dependent on the amount of dry mass available after collection. It is known that genotype and environmental variation occurs for sugarcane E.M. Sugarcane in the U.S., has ~34% total dry E.M. biomass [78] compared to 41% for some sugarcane varieties in South Africa [77]. Dried leaf residue contains approximately 36% cellulose, 21% hemicelluloses, and 16% lignin [79] and is similar to the composition of bagasse. The quantity and quality of leaves can also vary across the harvesting season in different countries. Donaldson et al. [80] in South Africa and Eggleston et al. [81] in Louisiana, U.S., both reported that the amounts of green leaves varied with season date while dried, brown leaves changed little with season.

5.2. Biochar from Extraneous Matter

Along with sugarcane bagasse, E.M. is another feasible feedstock for the manufacture of biochar. Varying amounts of soil, leaves, etc., present in E.M. will, however, produce heterogenous biochars of equally varying physico-chemical properties, which could present a challenge to their usability. Because E.M. is mainly composed of leaves, it is significantly less heavy than sugarcane bagasse, with approximately 4 to 5 times lower densities. This can become an issue during the field application of biochars due to wind losses, as well as during storage. Densification technologies can potentially address these challenges, although the moisture content would have to be reduced prior to pelletization to <20%. As with sugarcane bagasse, sugarcane molasses can be used as a binder in biochar manufacture from E.M. [62,63,66].

5.3. Collection and Storage of Sugarcane Extraneous Matter

If E.M. is to work as a reliable, sustainable, and economical biomass feedstock to the new biorefineries, then collection of excess E.M. in the field needs to be optimized, preferably after some solar drying to create greater dry mass [82]. Collection could include baling to increase the material's bulk density for transportation to the biorefinery. As an alternative, E.M. could be separated from the stalks at the biorefinery site if economical and effective separation technologies are in place such as dry cleaning before the sugarcane is shredded [83]. Alta Mogiana sugarcane factory in Sao Paulo State, Brazil has an E.M. dry cleaning system and the separated E.M. is burnt in the boilers. In 2014, NFR BioEnergy LLC, initiated an E.M. dry cleaning system of their own design in the USA, which is co-located with the Cora Texas sugarcane factory in White Castle, Louisiana [84]. NFR BioEnergy already has the technology to produce biochars from sugarcane bagasse and plan to also use separated E.M. as a feedstock. Their dry cleaning system incorporates a large drying tumbler that effectively removes E.M. from the cane. The collected E.M, together with excess bagasse, will be fed into a torrefaction or pyrolysis unit for the manufacture of biochars, although currently the unit is only pilot-plant size. In their process, synthesis gas will also be recovered and used to supply energy to the torrefier/pyrolysis unit. The objective is to create a new source of sustainable energy from using the pelleted biochar as fuel, but to also remove E.M. and allow for cleaner sugarcane to be processed at the factory. The removal of leaves and soil before it enters the factory will allow more sugar to be produced at a higher quality than was previously possible [85].

Energy generated from syngas released during torrefaction can be used to make the whole system "energy negative" and to help run additional equipment such as a pellet mill upstream from the torrefier [86]. However, questions still remain on how efficiently E.M. separation technologies perform while not removing valuable sucrose in stalks [74,87]. Moreover, in some cases separation at the biorefinery could also create excessively large piles that would have to be stored or rapidly utilized [74].

Just as for bagasse, if E.M. is to be a stable and sustainable biomass feedstock it will most likely be stored in piles at the biorefinery or near the harvest site. Unlike bagasse, however, very little is known on how piles of E.M. store and for how long. Eggleston et al. [88] reported on the deterioration of brown, dry (BL) and green leaves (GL), stripped from sugarcane whole-stalks, stored under simulated wet and dry conditions. The worst deterioration for both BL and GL, generally, occurred in the watered samples and when the humidity was highest. On deterioration, more soluble impurities were extracted from GL than BL. Only prolonged deterioration of BL caused a reduction of fiber biomass, and the fiber content of GL usually increased on deterioration because of loss of moisture. Because its shredded state created more surface area to absorb water, moisture was highest in deteriorated BL; this also allowed for more Leuconostoc mesenteroides bacteria to grow and form dextran and mannitol [88].

6. DEDICATED ENERGY FEEDSTOCKS FROM SUGAR CROPS

The manufacture of second generation biofuels is expected to utilize a much more diverse set of feedstocks compared to first generation biofuels [89], and dedicated energy crops represent one option. Dedicated energy crops offer high output per hectare with low inputs. They are also expected to grow on land less suitable for food production. Often, the new energy crops yield well under various stress conditions and have even been developed with advantageous processing characteristics [90]. Energy cane, beet and sorghum crops can be converted to second generation cellulosic fuel ethanol as well as power and bioelectricity. Processes to convert high fiber energy canes and beets into fuel ethanol are still under investigation as discussed in Section 4.2. The challenge is to develop energy crops with a suite of desirable physical and chemical traits while increasing biomass yields by a factor of two or more [91].

6.1. Energy Canes

Energy cane varieties are high-fiber and biomass clones of conventional sugarcane [92], and often have extended ratooning abilities. The goals of achieving energy cane with attributes for a biomass-based economy are readily achievable through existing sugarcane breeding programs [7]. In sugarcane breeding, more rapid genetic gain can occur for total biomass than sugar yield because growth does not have to be intentionally restricted during the life cycle of the crop, and a wider array of germplasm of potential value is available to the breeder once stringent standards for sucrose and fiber levels are relaxed. Breeding strategies for energy cane are classified as Type I and Type II. Type I are energy cane varieties close to conventional sugarcane but have lower sucrose content and thus higher fiber contents. Type II are energy cane varieties that have marginal sugar content and high fiber at such high levels (typically >16%) that raw sugar manufacturers consider unacceptable for processing [7]. Furthermore, the Type II energy canes would be utilized as biomass feedstock for the production of electricity and cellulosic biofuels [92]. For more information about Type I and Type II energy canes the reader is referred to Botha and Moore [7].

A few energy cane varieties have already been developed and released, for example in Louisiana, U.S. [93]. Three Type I high fiber sugarcane varieties (L 79-1002, HoCP 91-552, and Ho 00-961) released for commercial planting in 2007 produce dry biomass yields in excess of 25 tonnes/ha [93]. As marginal land to grow energy canes in Louisiana are mostly north and, therefore, colder during the winter, a major emphasis of the breeding program is to breed for cold tolerance.

6.2. Energy Beets

Many biofuel/bioproduct sugar feedstock producers, including U.S. large-scale sweet sorghum producers, are considering energy beets as a co-rotation crop to allow the year round production of sugar feedstocks. Energy beet is a non-edible, water efficient, selected hybrid of commercial sugar beet that has not been bred for sugar but for dry matter. The creation of higher biomass yields for energy beets have been made possible by using fodder beet germplasm as a parent in hybrids with sugar beet [94,95]. Biomass yield potential is dependent upon interception of solar radiation which gives beets grown in areas with long growing seasons a decided advantage. Energy beets are typically grown as winter crops (winter beets in the U.S. have a longer growing season and, therefore, a much higher yield potential), can grow on marginal lands, and also have higher fiber, glucose and fructose concentrations [96]. However, the glucose and fructose concentrations are still lower than in sweet sorghum, and the sucrose considerably higher. Overall, energy beets hybrids are comparatively new compared to conventional sugar beets, and very little processing research has been reported.

7. SUGAR AND SUGAR-BIOPRODUCT INDUSTRIES: PLATFORM CHEMICALS FROM SUGAR FEEDSTOCKS AND OTHER VALUE ADDED PRODUCTS FROM SUCROSE

7.1. Platform Chemicals from Sugar Feedstocks

Several years ago, Novozymes CEO Steen Riisgaard said "in a few years sugar will be the new oil" as sugar is a superb feedstock for the production of platform chemicals for the manufacture of a range of end-products, e.g., bioplastics, industrial solvents, and chemicals [97]. This statement still holds true today and structural bio-based materials, such as artificial spider silk "Spiber™ (Stockholm, Sweden)" which is stronger than steel and lighter than carbon fiber [98], can now be added to the portfolio of possible end-products. Furthermore, Koch et al. [99] recently stated that "a promising option for intermediates of sugar production, including syrup, is to ferment them and this is less expensive than pure sucrose". Scientists are also currently taking a closer look at natural sweeteners such as honey, maple syrup, molasses, and sweet sorghum syrup, as sugar replacements. This is because they contain, beyond sucrose, glucose, and fructose, other classes of bioactive compounds including complex carbohydrates, amino acids, and polyphenols that might impart health benefits [20].

Efforts in "green chemistry" have been ramped up to transform renewable crop biomass, e.g., from sugar crops, into the basic chemical ingredients that go into many everyday products [99,100]. Thanks to numerous years of work on engineered microbes and new catalysts, the reach of biobased chemicals into consumer items is expanding [101]. A host of biobased intermediates are at or near commercialization, and include raw materials for common polymers such as polyester, spandex, synthetic rubber, and nylon [101]. Such products are being manufactured by both start-up firms and industrial giants, with announcements of progress gaining in frequency and substance [101]. A large-scale shift to bio-based polymers, however, will depend on the availability and reliability of large quantities of sugar feedstock at competitive prices. Moreover, as Bomgardner [101] stated that commercializing a biobased polymer "requires sustained, parallel progress on several fronts including technology development, end-product verification, market demand, and robust business acumen." In 2004, DOE identified a set of biomass-derived compounds best suited to replace petroleum-derived chemicals [102], and Table 3 lists the current and projected commercialization of these products [101]. Only succinic acid, sorbitol, and xylitol productions are established on a large, commercial scale (Table 3). For example, DSM and Myriant are now manufacturing succinic acid in Italy [103] and the USA, respectively. It must be noted that some of the new, start-up companies that started out aiming for commodity biofuels and bioproducts have ended up manufacturing specialty compounds instead [104]. This is because selling in expensive markets helps pay the company bills when the start-up scale is small and its products are still costly to manufacture. Moreover, this strategy can allow companies to work their way into commodities [104].

Although there is no current, effective one-step method or multi-step methods for converting raw lignocellulose to finished products, progress is being made. Furthermore, increasing investments in the sugar-ethanol industry could facilitate the construction of the physical infrastructure, and associated technologies that could also be used for the production of bioproducts [97]. Biotechnology processes are particularly suited for the transformation of natural feedstock from sugar crops into the necessary sugars and building blocks of secondary bioproducts, and bioethanol itself can also be used as a platform chemical [97].

Table 3. Current and Projected Commercialization of Biobased Products from Sugar Feedstocks that were Identified in 2004 by the U.S. Department of Energy (DOE) and the most likely compounds to replace petroleum-derived chemicals. Adapted from Bomgardner [101].

Biobased Compound	Number of Carbon Atoms	Method of Manufacture	Key Uses and End-Products	Commercialization in 2014	Projected Commercialization in 2024
Succinic acid	4	Bacterial fermentation of glucose, chemical oxidation of 1,4-butanediol	Solvents, polyesters, poly-urethanes, nylon, food and beverage acid control, surfactants, adhesives, fabrics, inks, paints	Yes	Yes
2,5-Furanedicarboxylic acid	6	Chemical dehydration of glucose, oxidation of 5-HMF	Polymers such as nylon, plastic bottles/containers, carpet fiber	Soon	Yes
3-Hydroxypropionic acid	3	Bacterial fermentation of glucose	Polymers, carpet fiber, paints and adhesives, superabsorbent, contact lenses	No	Yes
Sorbitol	6	Hydrogenation of glucose from corn syrup, bacterial fermentation under development	Sweeteners, fuel ingredients, anti-freeze, water treatment	Yes	Yes
Xylitol	5	Hydrogenation of xylose, extraction from lingo-celluloses, bacterial fermentation under development	Sweeteners, cough drops and medicines, anti-freeze, new polyesters	Yes	Yes
Levulinic acid	5	Acid-catalyzed dehydration of sugars	Fuel ingredients, solvents, plastic bottles, polyesters, polyamides, pharmaceuticals, herbicides	No	Maybe
Itaconic acid	5	Fungal fermentation of glucose	Styrene-butadiene copolymers, rubber, plastics, paper	No	Maybe
3-Hydroxybutyrolactone	4	Multi-step chemical synthesis from starch	Solvents, synthetic intermediates for pharmaceuticals, new polymers	No	Maybe
Glutamic acid	5	Bacterial fermentation of glucose	Polyesters, nylon analogs, flavor enhancers, fabrics, plastics	No	Maybe

7.2. Value Added Products from Sucrose

For the sugar industry, value-added products from sucrose can increase the demand, value, and consumption of sucrose, as well as improve the industry's competitiveness. However, only a small percentage of the sugar produced in the world is used in non-food applications [105], which is unfortunate as much research effort and funds have been expanded on the identification and development of value-added products from sucrose. Part of the reason for such little impact of this research is that the scientists inventing the products have not fully considered the market, and do not have the business acumen to sell such products to industry [11]. More involvement by industry, particularly at the conception phase, would help to gain more impact [11]. Another explanation is that, historically, fossil-based oil has been a less expensive feedstock than sugars.

Sucrose is a good source for many value-added products because of its chemical and enzymatic reactivity. The basis for the reactivity of sucrose is the eight hydroxyl groups present on the molecule. Generally, the three primary hydroxyls have greater reactivity but they often prove a hindrance as they are difficult to react exclusively [106]. The synthesis of an enormous number of sucrose derivatives is possible; substitution with just one group type could theoretically give two hundred and fifty five different compounds. Moreover, the alcohol group can be derivatized to form esters, ethers, and substitution derivatives [106]. Sucrose can be readily degraded by acids, oxidizing agents, alkalis, and catalytic hydrogen to compounds of lower molecular weight. Sucrose is also an exceptional molecule for enzymatic synthesis reactions [105,107] to form, for example, oligosaccharides and polysaccharides [108].

8. VHP (VERY HIGH POL) AND VLC (VERY LOW COLOR) SUGAR PRODUCTION—A SUSTAINABLE TREND

In recent times there has been a world-wide trend to manufacture VHP and VLC raw sugars for supply to refineries, i.e., a trend of vertical integration from the field to the white sugar output. Furthermore, a concomitant trend exists to build refineries of the VHP/VLC cane raw sugar close to the consumption areas to satisfy the needs of the food industry. There is also a growing demand for exports of VHP and VVHP (very, very high pol) raw sugars, particularly from Brazil, mainly for overseas markets. Some refineries also want lower ash concentrations in the VHP/VLC sugar because: (a) some of the refined sugar will be manufactured into liquid sugar, which requires low ash; and (b) lower ash is needed for short, medium, and long term refinery strategies [4]. The supply of higher quality raw sugars is expected to create additional efficiencies at the new refineries, particularly at the early, energy-intensive affination stage. Some refineries such as Al Khalij Sugar in Dubai and Dangoto in Lagos are even eliminating the affination stage altogether. The higher quality raw sugars will also allow factory processors to gain premiums from the new refineries. Furthermore, manufacture of higher quality raw sugars at the factory where the energy source is renewable bagasse, will save fossil energy utilization by the refiners.

9. SUSTAINABILITY METRICS

Traditionally, chemical/food process and product development has focused on the assessment of economic criteria, but additional criteria for sustainability have become increasingly important and integrated into decision making processes [109]. Assessment tools, standards, and enhanced metrics to measure "green, greener, or greenest" have been and are continuing to be developed [110]. Many countries and regions have introduced policies or adopted standards to promote sustainable manufacture of biofuels and bioproducts and use, most prominently in the U.S. and European Union. Ecological or environmental sustainability, one of the three pillars of sustainability (Figure 2) can be examined using Life Cycle Assessment (LCA) [110]. LCA is an internationally recognized methodology for evaluating environmental performance of a product, process, or pathway along its partial or whole life cycle, which can be applied to new processes for converting sugar biomass [7]. Late in 2014, the World Business Council for Sustainable Development (WBCSD) in Geneva, Switzerland, published a guide titled "Life Cycle Metrics for Chemical Products" to help the chemical industry and associated stakeholders compare, on a common sense basis, the environmental footprint of chemical products [111]. This should allow chemical sector companies, including those using sugar feedstocks, to communicate with a common language to companies downstream, and help scaling up solutions [111]. Current operating chemical companies such as BASF, Amcor, and Dow Chemical are benefiting from using sustainability metrics to meet the challenge of improving processes and products, and have already provided insights on opportunities and progress [112].

Rein [113] specifically reported on measuring and monitoring sustainability in the sugar and sugar-ethanol industries. The Better

Sugarcane Initiative (BSI) is a collaboration of sugar retailers, investors, traders, producers, and NGOs who are committed to sustainable sugar production by establishing principles and criteria that can be applied to sugarcane [113]. In the sugarcane industry, Brazil has been the most active in embracing and reporting sustainability performance, mostly because they export to other countries [113], although broader sustainability reporting relating to social and other environmental issues have been less of a focus in Brazil. For the future sustainability of the sugar and sugar-bioproduct industries, there is also a need for new analytical methods in biofuels and bioproducts manufacture as well as for grower payment systems with the new feedstock crops [4].

10. OVERALL OUTLOOK

Biomass from sugar crops is expected to become a significant part of the world-wide shift from a fossil-fuel based economy to one that is biobased [7]. This shift, however, is still mostly in the early pioneering stage. Key criteria in the establishment of commercial biobased plants are cost of the facility, availability of reliable quality feedstocks at low cost, financing, and the policy environment [103]. Concomitantly, in many areas of the world, there is a progressive diversification of the sugar industry into "sugar and sugar-bioproduct industries" that are deeply involved in the maximization of the sugar crop biomass. This includes Europe where many beet factories have diversified to produce ethanol as well as sugar from sugar beets. An excellent example of a sugar company diversifying into biobased chemicals while obtaining social, environmental, and financial sustainability gains, is Godavari in India [114]. Starting with sugarcane, Godavari manufactures refined sugar, ethanol, and chemicals including ethyl acetate, crotonaldehyde, 1,3-butanediol, and even flavor and fragrance ingredients [114]. From bagasse, electricity is cogenerated and used to power the sugar and chemicals factories and plants. Godavari's biobased chemical are produced so efficiently that they are able to compete pricewise against the same chemical produced from fossil sources [114]. It is expected that more "sugar" and "sugar-bioproduct" companies like Godavari, will become more and more eager to become greener [110] as they realize that they can reduce pollution and increase profits simultaneously [115]. Moreover, such companies will want to be able to select greener starting materials and use cleaner chemical processes to make environmentally preferred products [110], and to appease consumers who are becoming increasingly aware of sustainability issues [116]. Consumers are, in fact, holding companies to higher standards and asking them to demonstrate the wider impact of their operations [116].

Overall, the sustainability of the sugar and sugar-bioproduct industries should be viewed as a continuous improvement journey [1], and behavior change and education will also be linchpins in effective sustainability programs.

ACKNOWLEDGMENTS

Mention of trade names or commercial products in this article is solely for the purpose of providing specific information and does not imply recommendation or endorsement by the U.S. Department of Agriculture. USDA is an equal opportunity provider and employer.

AUTHOR CONTRIBUTIONS

Gillian Eggleston led the research on sugar processing, the effect of sugarcane extraneous mass on processing and how it deteriorates on storage. She also conducted the review of literature on sugar processing, use of sugar crops as a biomass source. Isabel Lima led and conducted the research on biochar production from bagasse and extraneous matter, and also conducted the review of literature on these topics.

CONFLICTS OF INTEREST

The authors declare no conflict of interest.

REFERENCES

1. Higgins, K.T. Tastes great, less carbon. Food Eng. 2009, 9, 51–58.
2. Brundtland, G.H. Brundtland Report; Oxford University Press: Oxford, UK, 1987.

3. Abraham, M.A.; Clark, J.; Winterton, N. Introduction. In Green Chemistry Metrics; Lapkin, A., Constable, D., Eds.; Wiley & Sons: Oxford, UK, 2008; pp. 1–26.

4. Eggleston, G. Future sustainability of the sugar and sugar-ethanol industries. In Sustainability of the Sugar and Sugar-Ethanol Industries; ACS Symposium Series 1058; Eggleston, G., Ed.; Oxford University Press: Oxford, UK, 2010; pp. 1–19.

5. Chen, P.; Wan, Y.; Wang, X.; Cheng, Y.; Liu, Y.; Lin, X.; Ruan, R. Bioenergy industry status and prospects. In Industrial Crops and Uses; Singh, B.P., Ed.; CABI: Wallingford, UK, 2010; pp. 21–34.

6. Jiang, D.; Zhuang, D.; Huang, Y. Crop residues as an energy feedstock. Availability and sustainability. In Sustainable Bioenergy Production; Wang, L., Ed.; CRC Press: Boca Raton, FL, USA, 2014; pp. 223–235.

7. Botha, F.C.; Moore, P.H. Biomass and bionergy. In Sugarcane: Physiology, Biochemistry, and Functional Biology; Botha, F.C., Moore, P.H., Eds.; Wiley & Sons: Hoboken, NJ, USA, 2014; pp. 521–540.

8. Han, F.X.; King, R.L.; Lindner, J.S.; Yu, T.Y.; Durbha, S.S.; Younan, N.H.; Monts, D.L.; Su, Y.; Luthe, J.C.; Plodinec, M.J. Nutrient fertilizer requirements for sustainable biomass supply to meet US bioenergy goal. Biomass Bioenergy 2011, 35, 253–262.

9. Van Dam, J.; Junginger, M. Striving to further harmonization of sustainability criteria for bioenergy in Europe: Recommendations from a stakeholder questionnaire. Energy Policy 2011, 39, 4051–4066.

10. Ottewell, S. Europe targets biofuel sustainability. Chemical Processing, 20 November 2014, 58.

11. Eggleston, G. Sucrose and related oligosaccharides. In Glycoscience; Fraser-Reid, B., Tatsuta, K., Thiem, J., Eds.; Springer-Verlag: Berlin, Germany, 2008; pp. 1163–1182.

12. Van der Poel, P.W.; Schiweck, H.; Schwartz, T. Sugar Technology. Beet and Cane Sugar Manufacture; Verlag Dr. Albert Bartens KG: Berlin, Germany, 1998.

13. Chen, J.C.P.; Chou, C.C. Cane Sugar Handbook, 12th ed.; John Wiley and Sons, Inc.: New York, NY, USA, 1993.

14. Hugot, E. Handbook of Cane Sugar Engineering, 2nd ed.; Elsevier: Amsterdam, The Netherlands, 1972.

15. McGinnis, R.A. Beet-Sugar Technology, 3rd ed.; Beet Sugar Development Foundation: Fort Collins, CO, USA, 1982.

16. Ragagopal, D.S.; Sexton, G.; Hochman, D.; Roland-Holst, D.; Zilberman, D. Model estimates food-versus-biofuel trade-off. Calif. Agric. 2009, 63, 199–201.

17. RFA: Renewable Fuels Association. World Fuel Ethanol Production. Available online: http://ethanolrfa.org/pages/World-Fuel-Ethanol-Production (accessed on 24 August 2015).

18. F.O. Licht World Ethanol and Biofuel Report. Available online: https://store.agra-net.com/ (accessed on 24 August 2015).

19. U.S. Environmental Protection Agency. Greenhouse Gas Reduction Thresholds. 2010. Available online: http://www.epa.gov/OMS/renewablefuels/420f10007.htm#7 (accessed on 28 June 2015).

20. Ritter, S.K. Green chemistry awards. C&E News, 20 October 2014, 32–34.

21. Blakarishnan, M.; Sacia, E.R.; Sreekumar, S.; Gunbas, G.; Gokhale, A.A.; Scown, C.D.; Dean Toste, F.; Bell, A.T. Novel pathways for fuels and lubricants from biomass optimized using life-cycle greenhouse gas assessment. Proc. Natl. Acad. Sci. USA 2015, 112, 7645–7649.

22. Koninckx, J. Biorefinery beckons: Iowa plant will produce ethanol from corn stover. Chemical Processing, 3 January 2013, 38–39.

23. Kenney, K.L.; Smith, W.A.; Gresham, G.L.; Westover, T.L. Understanding biomass feedstock variability. Biofuels 2013, 4, 111–127.

24. Eggleston, G.; Cole, M.; Andrzejewski, B. New commercially viable processing technologies for the production of sugar feedstocks from sweet sorghum (Sorghum bicolor L. Moench) for biofuel and bioproducts manufacture. Sugar Tech. 2013, 15, 232–249.

25. Smith, B.A.; Romo, R.V.; Smith, R.A.; de la Cruz, R.A.; Lime, B.J. Production of raw sugar from sorghum juices. Sugar J. 1973, 35, 22–27.

26. Rooney, W.L.; Blumenthan, J.; Bean, B.; Mullet, J.E. Designing sorghum as a dedicated bioenergy feedstock. Biofuels Bioprod. Biorefin. 2007, 1, 147–157.

27. Tew, T.L.; Cobill, R. Chapter 9: Genetic improvement of sugarcane (Saccharum spp.) as an energy crop. In Genetic Improvement of Bioenergy Crops; Vermerris, W., Ed.; Springer Science: Berlin, Germany, 2008; pp. 249–272.

28. Eggleston, G.; Heckemeyer, M.; St. Cyr, E.; Wartelle, L. Case study: Commercialization of sweet sorghum juice clarification for large-scale syrup manufacture. Sugar Tech. 2015.

29. Fanjul, J.; Okeelanta Corporation, South Bay, FL, USA. Personal communication, 2014.

30. Serna-Saldivar, S.O.; Rooney, W.L. Chapter 8: Production and supply logistics of sweet sorghum as an energy feedstock. In Sustainable Bioenergy Production; Wang, L., Ed.; CRC Press: Boca Raton, FL, USA, 2014; pp. 193–212.

31. Dixon, T.F.; Bullock, G.E. Bioenergy outlook for Australia—For sugar not so sweet. Zuckerindustrie 2004, 129, 233–236.

32. Caio Filho, L. Bioelectricity: An increasing source of income in the sector. In F. O. Licht 2010 Yearbook; F. O. Licht: London, UK, 2010; pp. D13–D15.

33. Bagasse-based cogen sector expanding. Available online: https://www.internationalsugarjournal.com/brazil-bagasse-based-cogen-sector-expanding (accessed on 24 August 2015).

34. Castro, A.G.; Quiros, E. Cogeneration of energy from sugarcane bagasse in Costa Rica: The case of El Viejo factory. Intern. Sugar J. 1991, 93, 184–190.

35. Goldemberg, J.; Coelho, S.T.; Nastari, P.M.; Guardabasi, P. Chapter 9: Production and supply logistics of sugarcane as an energy feedstock. In Sustainable Bioenergy Production; Wang, L., Ed.; CRC Press: Boca Raton, FL, USA, 2014; pp. 213–222.

36. Fischer-Tropsch (FT). Available online: www.velocys.com (accessed on 24 August 2015).

37. Purchase, B.S.; Rosettenstein, S.; Bezuidenhoudt, D.V. Challenges and potential solutions for storage of large quantitites of bagasee for power generation. Intern. Sugar J. 2014, 116, 592–602.

38. Solomon, B.D.; Barnes, J.R.; Halvorsen, K.E. Grain and cellulosic ethanol: History, economics, and energy policy. Biomass Bioenergy 2007, 31, 416–425.

39. Petrou, E.C.; Pappis, C.P. Biofuels: A survey of pros and cons. Energy Fuels 2009, 23, 1055–1066.

40. Brennan, L.; Owende, P. Biofuels from microalgae—A review of technologies for production, process, and extraction of biofuels and co-products. Renew. Sustain. Energy Rev. 2010, 14, 557–577.

41. Dantas, G.A.; Legey, L.F.L.; Mazzone, A. Energy from sugarcane bagasse in Brazil: An assessment of the productivity and cost of different technological routes. Renew. Sustain. Energy Rev. 2013, 21, 356–364.

42. Yaman, S. Pyrolysis of biomass to produce fuels and chemical feedstocks. Energy Conv. Manag. 2004, 45, 651–671.

43. Spokas, K.A.; Cantrell, K.B.; Novak, J.M.; Archer, D.W.; Ippolito, J.A.; Collins, H.P.; Boateng, A.A.; Lima, I.M.; Lamb, M.C.; McAloon, A.J.; et al. Biochar: A synthesis of its agronomic impact beyond carbon sequestration. J. Environ. Qual. 2011, 41, 973–989.

44. Bapat, H.D.; Manahan, S.E. Chemchar gasification of hazardous wastes and mixed wastes on a biochar matrix. Available online: http://www.biochar-international.org/node/ 915 (accessed on 24 August 2015).

45. Mishra, B.K.; Ramakrishnan, P.S. Slash and burn agriculture at higher elevations in North-Eastern India. I. Sediment, water and nutrient losses. Agric. Ecosyst. Environ. 1983, 9, 69–82.

46. German, L.A. Historical contingencies in the coevolution of environment and livelihood: Contributions to the debate on Amazonian Black Earth. Geoderma 2003, 111, 307–331.

47. Mann, C.C. 1491: New revelations of the Americas before Columbus; Vintage and Anchor Books: New York, NY, USA, 2005.

48. Bustin, R.M.; Guo, Y. Abrupt changes (jumps) in reflectance values and chemical compositions of artificial charcoals and inertinite in coals. Int. J. Coal Geol. 1999, 38, 237–260.

49. DeLuca, T.H.; MacKenzie, M.D.; Gundale, M.J. Biochar effects on soil nutrient transformations. In Biochar for Environmental Management: Science and Technology; Lehmann, J., Joseph, S., Eds.; Earthscan: London, UK, 2009; pp. 251–270.

50. Lehmann, J.; Rilig, M.C.; Thies, J.; Masiello, C.A.; Hockaday, W.C.; Crowley, D. Biochar effects on soil biota—A review. Soil Biol. Biochem. 2011, 43, 1812–1836.

51. Lehmann, J.; Gaunt, J.; Rondon, M. Bio-char sequestration in terrestrial ecosystems—A review. Mitig. Adapt. Strat. Glob. Chang. 2006, 11, 403–427.

52. Lima, I.M.; White, P.; Bigner, R.; Klasson, K.T. Economic products derived from sugarcane trash and bagasse. Available online: https://www.sugarjournal.com (accessed on 24 August 2015).

53. Zandersons, J.; Gravitis, J.; Kokorevics, A.; Zhurinsh, A.; Bikovens, O.; Tardenaka, A.; Spince, B. Studies of the Brazilian sugarcane bagasse carbonization process and product properties. Biomass Bioenergy 1999, 17, 209–219.

54. Pendyal, B.; Johns, M.M.; Marshall, W.E.; Ahmedna, M.; Rao, R.M. The effect of binders and agricultural by-products on physical and chemical properties of granular activated carbons. Bioresour. Technol. 1999, 68, 247–254.

55. Erlich, C.; Ohman, M.; Björnborn, E.; Fransson, T.H. Thermochemical characteristics of sugarcane bagasse pellets. Fuel 2005, 84, 569–575.

56. Cordero, T.; Marquez, F.; Rodriguez-Mirasol, J.; Rodriguez, J.J. Predicting heating values of lignocellulosics and carbonaceous materials from proximate analysis. Fuel 2001, 80, 1567–1571.

57. Channiwala, S.A.; Parikh, P.P. A unified correlation for estimating HHV of solid, liquid and gaseous fuels. Fuel 2002, 81, 1051–1063.

58. Garcia-Perez, M.; Chaala, A.; Roy, C. Vacuum pyrolysis of sugarcane bagasse. J. Anal. Appl. Pyrol. 2002, 65, 111–136.

59. Parikh, J.; Channiwala, S.A.; Ghosal, G.K. A correlation for calculating HHV from proximate analysis of solid fuels. Fuel 2005, 84, 487–494.

60. Munir, S.; Daood, S.S.; Nimmo, W.; Cunliffe, A.M.; Gibbs, B.M. Thermal analysis and devolatization kinetics of cotton stalk, sugarcane bagasse, and shea meal under nitrogen and air atmospheres. Bioresour. Technol. 2009, 100, 1413–1418.

61. Katyal, S.; Thambimuthu, K.; Valix, M. Carbonisation of bagasse in a fixed bed reactor: Influence of process variables on char yield and characteristics. Renew. Energy 2003, 28, 713–725.

62. Ahmedna, M.; Marshall, W.E.; Rao, R.M. Production of granular activated carbons from select agricultural by-products and evaluation of their physical, chemical and adsorption properties. Bioresour. Technol. 2000, 71, 113–123.

63. Ahmedna, M.; Marshall, W.E.; Rao, R.M. Surface properties of granular activated carbons from agricultural by-products and their effects on raw sugar decolorization. Bioresour. Technol. 2000, 71, 103–112.

64. Ng, C.; Bansode, R.R.; Marshall, W.E.; Losso, J.N.; Rao, R.M. Process description and product cost to manufacture sugarcane bagasse-based granular activated carbon. Intern. Sugar J. 2002, 104, 401–408.

65. Ng, C.; Marshall, W.; Rao, R.M.; Bansode, R.R.; Losso, J.N.; Portier, R.J. Granular Activated Carbons from Agricultural By-products: Process Description and Estimated Cost of Production, Bulletin No. 881; Louisiana State University AgCenter Research & Extension: Baton Rouge, LA, USA, 2003.

66. Johns, M.M.; Toles, C.A.; Marshall, W.E. Activated Carbons from Low-density Agricultural Waste. U.S. Patent 6,537,947, 25 March 2003.

67. Lehmann, J. Bio-energy in the black. Front. Ecol. Environ. 2007, 5, 381–387.

68. Bracmort, K.S. Biochar: Examination of an Emerging Concept to Mitigate Climate Change; CRC Report for Congress R40186; Congressional Research Service: Washington, DC, USA, 2009.

69. The "Bagatex-20" Process for Rapid Drying of Baled Bagasse by Accelerated and Controlled Fermentation, Using a Biochemical Catalyst; Groupe Sucres et Denrees: Paris, France, 1986; p. 49.

70. Springer, E.L. Should Whole-tree Chips for Fuel be Dried before Storage? Forest Products Laboratory Research Note FPL-0241; United States Department of Agriculture: Washington, DC, USA, 1979.

71. Wright, M.; Lima, I.M.; Bigner, R. Microbial and physicochemical properties of sugarcane bagasse for potential conversion to value-added products. Sugar J. 2015, 78. Article 31.

72. Parikh, J.; Channiwala, S.A.; Ghosal, G.K. A correlation for calculating elemental composition from proximate analysis of biomass materials. Fuel 2007, 86, 1710–1719.

73. Munoz-Arboleda, F.; Quintero-Duran, R. Trash management after green cane harvesting and its effect on productivity and soil respiration. Intern. Sugar J. 2011, 113, 804–808.

74. Viator, R.P.; Johnson, R.M.; Richard, E.P., Jr. Challenges of post-harvest residue retention management in the Louisiana sugarcane industry. Proc. ISSCT Congr. 2005, 25, 238–244.

75. Singh, P.; Suman, A.; Tiwari, P.; Arya, N.; Gaur, A.; Shrivatava, A.K. Biological pretreatment of sugarcane trash for its conversion to fermentable sugars. World J. Microbiol. Biotechnol. 2008, 24, 667–673.

76. Ripoli, M.L.C.; Ripoli, T.C.C.; Villa Nova, N.A. Sugarcane crop residues and bagasse for cogeneration in Brazil. Am. Soc. Agric. Biol. Eng. Pap. 2007.

77. Muir, B.; Eggleston, G. Factory trials to determine how sugarcane trash impacts downstream processing including affinated sugar production. Proc. Sugar Ind. Technol. Meet. 2009, 68, 24–48.

78. Eggleston, G.; Grisham, M.; Tew, T.; Triche, R.; Antoine, A. Potential biomass quantity and processing quality of trash tissues by different US sugarcane varieties. Intern. Sugar J. 2009, 111, 108–118.

79. Deepchand, K. Chapter 4: The sugarcane crop for the sustainable production of sugar and other cane derived products in Mauritius. In Sustainability of the Sugar and Sugar-Ethanol Industries; ACS Symposium Series; Eggleston, G., Ed.; Oxford University Press: Oxford, UK, 2011; pp. 53–72.

80. Donaldson, R.A.; Redshaw, K.A.; Rhodes, R.; van Antwerpan, R. Season effects on productivity of some commercial South African sugarcane cultivars. Part II: Trash production. Proc. S. Afr. Technol. Assoc. 2008, 81, 528–538.

81. Eggleston, G.; Viator, R.; Gateuil, A.; Fenger, J.; White, P.; Jackson, W.; Waguespack, H., Jr.; White, P.; Blackwelder, N. Effects of seasonal variations of sugarcane stalk and extraneous matter quantity and quality on recoverable sugar, starch, and fiber (Part I). Intern. Sugar J. 2013, 115, 477–487.

82. Purchase, B.S.; Wynne, A.T.; Meyer, E.; van Antwerpen, R. Is there profit in cane trash?—Another dimension to the assessment of trashing versus burning. Proc. S. Afr. Technol. Assoc. 2008, 81, 86–99.

83. Schembri, M.G.; Hobson, P.A.; Paddock, R. The development of a prototype factory-based trash separation plant. Proc. Aust. Sugar Cane Technol. 2002, 24, 12–18.

84. Young, R.D. NFR Bioenergy to start $312 million sugar refinery project across South Louisiana. Available online: www.nola.com/business/baton-rouge/index.ssf/2014/09/nfr_bioenergy_to_start_312 (accessed on 24 August 2015).

85. Eggleston, G.; Viator, R.; Gateuil, A.; Fenger, J.; White, P.; Jackson, W.; Waguespack, H., Jr.; White, P.; Blackwelder, N. Seasonal variations of sugarcane stalk and extraneous matter on pH, color and ash as they affect the production of high quality raw sugars (Part II). Intern. Sugar J. 2013, 117, 646–653.

86. NFR BioEnergy LLC; White Castle, LA, USA. Personal communication, 2015.

87. Viator, R.P.; Richard, E.P., Jr.; Viator, B.J.; Jackson, W.; Waguespack, H.; Birkett, H. Sugarcane chopper harvester extractor fan and ground speed effects on sugar yield, cane quality, and field losses. Appl. Eng. Agric. 2006, 23, 31–34.

88. Eggleston, G.; Klich, M.; Antoine, A.; Beltz, S.; Viator, R. Brown and green sugarcane leaves as potential biomass: How they deteriorate under dry and wet storage conditions. Ind. Crops Prod. 2014, 57, 69–81.

89. Salassi, M.; Brown, K.; Hilbun, B.; Deliberto, M.; Gravois, K.; Mark, T.; Falconer, L. Farm-scale cost of producing perennial energy cane as a biofuel feedstock. Bioenergy Res. 2014, 7, 609–619.

90. Zhou, M.; Kimbeng, C.A.; Eggleston, G.; Viator, R.; Hale, A.; Gravois, K. Issues of starch in sugarcane processing and prospects of breeding for low starch content in sugarcane. Sugar Cane Int. 2008, 26, 3–13.

91. Ragauskas, A.J.; Williams, C.K.; Davison, B.H.; Britovsek, G.; Cairney, J.; Eckert, C.A.; Frederick, W.J., Jr.; Hallett, J.P.; Leak, D.J.; Liotta, C.L.; et al. The path forward for biofuels and biomaterials. Science 2007, 311, 484–489.

92. Alexander, A.G. The Energy Cane Alternative; Elsevier: New York, NY, USA, 1985.

93. Anonymous. Release of three high fiber sugarcane varieties: L 79-1002, Ho 00-961, and HoCP 91-552. Sugar Bull. 2007, 85, 21–26.

94. Doney, D.L.; Theurer, J.C. Potential of breeding for ethanol fuel in sugarbeet. Crop Sci. 1984, 24, 255–257.

95. Theurer, J.C.; Doney, D.L.; Smith, G.A.; Lewellen, R.T.; Hogaboam, G.J.; Bugbee, W.M.; Gallian, J.J. Potential ethanol production from sugar beeet and fodder beet. Crop Sci. 1987, 27, 1034–1040.

96. Libsack, S.; EnerGynetics, Minneapolis-St. Paul, MN, USA. Personal communication, 2014.

97. Buttazoni, M. GHG Emission Reductions with Industrial Biotechnology; World Wide Fund for a Living Planet Report; WWF: Copenhagen, Denmark, 2008.

98. Borghino, D. Artificial "Spiber" Silk is Tougher than Kevlar. Available online: http://www.gizmag.com/spiber-sythetic-sil/28267 (accessed on 24 August 2013).

99. Koch, T.J.; Venus, J.; Bruhns, M. Sugar beet syrups in lactic acid fermentation—Part I. Zuckerindustrie 2014, 139, 495–502.

100. Jacoby, M. Chemicals from the garden. C&E News, 6 July 2009, 26–28.

101. Bomgardner, M.M. Biobased polymers. C&E News, 27 October 2014, 10–14.

102. Werpy, T.; Petersen, G. Top Value Added Chemicals from Biomass: Results from Screening for Potential Candidates from Sugars and Synthesis Gas. Available online: www1.eere.energy.gov/biomass/pdfs/35523.pdf (accessed on 24 August 2004).

103. Scott, A. Biomaterial blues. C&E News, 20 October 2014, 20–21.

104. Tullo, A.H. Endgame nears for Metabolix. C&E News, 25 May 2015, 26–27.

105. Godshall, M.A. Future directions for the sugar industry. Intern. Sugar J. 2001, 103, 378–384.

106. Walford, S.N.; Morel du Boil, P.G. A survey of value addition in the sugar industry. Proc. S. Afr. Technol. Assoc. 2006, 80, 39–611.

107. Eggleston, G.; Salassi, M.; Richard, E.; Birkett, H. Sustainability of the sugar industry: Future value addition from sugarcane. Intern. Sugar J. 2007, 109, 415–432.

108. Godshall, M.A. Value-added products for a sustainable sugar industry. In Sustainability of the Sugar and Sugar-Ethanol Industries; ACS Symposium Series; Eggleston, G., Ed.; Oxford University Press: Oxford, UK, 2010; pp. 253–268.

109. Kralisch, D. Applications of LCA in process development. In Green Chemistry Metrics; Lapkin, A., Constable, D., Eds.; Wiley & Sons: Oxford, UK, 2008; pp. 248–271.

110. Ritter, S.K. Common ground for going green. Chem. Eng. News, 10 May 2010, 38–41.

111. Rosenzweig, M. Group offers sustainability guide. Available online: http://www.chemicalprocessing.com (accessed on 24 August 2015).

112. Ottewell, S. Sustainability Efforts measure up. Chemical Processing, 27 April 2015, 20–25.

113. Rein, P. Measuring and monitoring sustainability in the sugar and sugar-ethanol industry. In Sustainability of the Sugar and Sugar-Ethanol Industries; ACS Symposium Series 1058; Eggleston, G., Ed.; Oxford University Press: Oxford, UK, 2010; pp. 23–35.

114. Trembley, J.-F. Sustainability yields sweet success. C&E News, 1 June 2015, 18–19.

115. Hart, S.L. Beyond greening strategies for a sustainable world. In Harvard Business Review on Green Business Strategy; Harvard Business School Publishing Corporation: Watertoen, MA, USA, 2007; pp. 99–123.

116. Leale, L.L. Industry is reducing its ecological footprint. Food Quality and Safety, 15 June 2015, 20–25.

8

CHAPTER

SAVING THE PLANET'S CLIMATE OR WATER RESOURCES? THE TRADE-OFF BETWEEN CARBON AND WATER FOOTPRINTS OF EUROPEAN BIOFUELS

Markus Berger [1,*], Stephan Pfister [2], Vanessa Bach [1] and Matthias Finkbeiner [1]

[1] *Chair of Sustainable Engineering, Department of Environmental Technology, Technische Universität Berlin, 10623 Berlin, Germany*

[2] *ETH Zürich, Institute of Environmental Engineering, 8093 Zurich, Switzerland*

ABSTRACT

Little information regarding the global water footprint of biofuels consumed in Europe is available. Therefore, the ultimate origin of feedstock underlying European biodiesel and bioethanol consumption was investigated and combined with the irrigation requirements of different crops in different countries. A (blue) water consumption of 1.9 m^3 in 12 countries per GJ of European biodiesel and 3.3 m^3 in 23 countries per GJ of bioethanol was determined. Even though this represents an increase by a factor of 60 and 40 compared to fossil diesel and gasoline, these figures are low compared to global average data. The assessment of local consequences has shown that the irrigation of sunflower seed in Spain causes 50% of the impacts resulting from biodiesel—even though it constitutes only 0.9% of the feedstock. In case of bioethanol production, the irrigation of sugar cane in Egypt, which constitutes only 0.7% of the underlying feedstock, causes 20% of the impacts. In a case study on passenger cars, it was shown that biofuels can reduce the global warming potential by circa 50% along the product life cycle. However, the price of this improvement is an approximate 19 times increased water consumption, and resulting local impacts are even more severe.

Keywords: water footprint; carbon footprint; biofuels; cars; transport

1. INTRODUCTION

Taking into account the European Union's targets to reduce both CO_2 emissions and its dependency on fossil fuels, biofuels such as biodiesel and bioethanol gained increasing attention during the last years. Fixed in the Renewable Energy Directive (RED) [1], the European Commission committed itself to reach a share of 10% of energy in road transport coming from renewable sources in each of the member states by 2020. Even though this includes hydrogen or electricity produced from renewable sources as well, biofuels are expected to be the most relevant fuels to reach this goal [2].

While the RED demands certain sustainability criteria to be met, such as the ban of cultivation of energy plants on land with high biodiversity and high carbon stocks, the environmental sustainability of biofuels is debated. Biofuels are not 'carbon neutral' as the cultivation and harvesting of energy plants as well as the production of the biofuels requires fertilizers, agricultural machinery, processing equipment and chemicals which all cause fossil CO_2 emissions during their production. Taking into account these aspects, recent life cycle assessment [3] and carbon footprint [4] studies determined a reduction potential in the global warming potential (GWP) of up to ~50% for biodiesel (soy bean, US) and ~65% for bioethanol (sugar cane, Brazil) compared to their fossil alternatives [5]. Even though this study shows that the minimum GWP reduction of currently 35% (50% in 2017, 60% in 2018) required by the RED is met in many cases, some scientists predict lower GWP reduction rates when considering GHG emissions resulting from the indirect land use effect [6]. However, due to the lack of scientific robustness, international standards do not require the inclusion of these indirect effects [7].

Despite these discussions, biofuels tend to show a more or less distinct advantage when it comes to the reduction of GHG emission. However, in comparisons to fossil fuels, it is often forgotten that the production of biodiesel and bioethanol can require a lot of water, mainly in the cultivation of crops. Studies of Gerbens-Leenes and colleagues [8] have shown that up to 11,636 L of “green”, “blue”, and “gray” water is needed to produce 1 L of biodiesel from jatropha and the production of 1 L of bioethanol from sorghum can require up to 4254 L of water, even though these plants are not used as a source of European biofuel production. Taking into account the facts that water consumption grew twice as fast as the world’s population during the past century and that more than 1 billion people live in water stressed areas, the neglecting of this aspect is a severe deficit in the discussion of international biofuel strategies. In a nutshell, this trade-off can be summarized as: Which water footprint increase do we need to accept for a carbon footprint decrease?

In order to tackle this question, this paper determines the water footprint of the European biodiesel and bioethanol consumption mixes and compares them to their fossil alternatives diesel and gasoline. On the inventory level, the volumes of blue water consumed in the global production of the underlying crops as well as in the crude oil production and refinery steps are investigated. As this study represents a water availability footprint according to ISO 14064 [9], green and gray water consumption are excluded here but can be reviewed in the works of Gerbens-Leenes and Hoekstra [10]. In order to allow for comparability, a functional unit of 1 GJ of fuel has been chosen, which equates to 31 L of biodiesel, 47 L of bioethanol, 28 L of diesel and 33 L of gasoline. Since the determination of water consumption volumes represents the first step in a water footprint analysis only [9], resulting local impacts are determined using recently developed impact assessment models. Finally, the influence of biofuels on the carbon and water footprint of a car is estimated taking into account the entire product life cycle.

2. METHODOLOGY

2.1. Water Inventory

Taking into account the findings of previous studies accomplished by Mekonnen and Hoekstra [11], it was assumed that relevant water consumption in the production of both biodiesel and bioethanol occurs only in the cultivation of the underlying crops resulting from irrigation. Here it should be noted, that water consumption refers to the fraction of total withdrawal which is not returned to the originating catchment area, mainly due to evapotranspiration [12]. As irrigation water needs highly depend on the crop and the location it is grown, the shares and origins of specific crops underlying the European consumption of biodiesel and ethanol need to be determined. Based on a study of Ecofys [13], Gerasimchuk [2] determined the ultimate origin of feedstock underlying the European biodiesel and bioethanol consumption in 2010 (Table 1 and Table 2). The underlying surveys also acknowledge triangular trade, e.g., the import of biodiesel from Argentina and Indonesia via the US. Moreover, the figures presented relate to the ultimate origin of feedstock. Thus, it is considered that the underlying crops used for the production of biofuels in a certain country can come from other countries as well. In order to provide a higher spatial resolution, total amounts derived from the European Union have been disaggregated based on the production shares of the individual crops on member state level derived from FAO data [14].

Subsequently, the country specific production volumes underlying European biodiesel (Table 1) and bioethanol (Table 2) consumption mixes have been combined with information regarding their country specific irrigation requirements. For this, the blue water footprint of biodiesel (Table 3) and bioethanol (Table 4) produced from different crops in different countries derived from the WaterStat Database ([11] reference year1996–2005) has been used.

The average water consumption per GJ of European biofuel has been calculated by dividing the total volume of water consumed to irrigate the underlying crops by the total amount of biodiesel (Table 1) and bioethanol (Table 2).

The water consumption per GJ of fossil diesel and gasoline has been determined with the GaBi 6 database [15] by using the water consumption figures of the EU27 fuel mixes (reference year 2013).

In order to assess local consequences of water consumption, the volumetric inventory information is not sufficient and spatial information is the minimum requirement for all impact assessment methods [16]. In case of biofuels, spatial data is available as the inventories have been determined based on the underlying feedstock and its origin. For fossil fuels, the location of water consumption was determined in a top-down regionalization approach introduced by Berger and colleagues [17,18]. By means of the GaBi 6 database, the total water consumption of diesel and gasoline has been subdivided into the shares of water consumed at the crude oil production and at the refineries. These shares have been allocated to different countries based on the European crude oil consumption mix and on the location of European refineries [15].

In addition to the absolute water consumption volumes, the spatially explicit water inventories show the location of water consumption along the supply chains of biodiesel, bioethanol, fossil diesel, and gasoline.

Table 1. Origin of crops underlying the European biodiesel consumption mix in 2010 (10^3 GJ) [2,14]; gray color indicates exclusion from water footprint calculation as the origin is unknown or the feedstock is a waste.

	Rapeseed	Soybean	Oil palm	Sunflower	Tallow	RVO	Other	Total
European Union	171,583	3643	209	18,590	6657	49,490	126	250,299
Austria	-	288	-	-	-	-	-	392
Bulgaria	-	-	-	4043	-	-	-	4043
Croatia	-	387	-	273	-	-	-	659
Czech Republic	13,050	-	-	-	-	-	-	13,050
Denmark	6203	-	-	-	-	-	-	6203
France	39,516	383	-	3303	-	-	-	43,203
Germany	52,305	-	-	-	-	-	-	52,409
Greece	-	-	-	703	-	-	-	703
Hungary	4769	285	-	3068	-	-	-	8122
Italy	-	1641	-	467	-	-	-	2108
Lithuania	4962	-	-	-	-	-	-	4962
Poland	24,213	-	-	-	-	-	-	24,213
Romania	6349	521	-	4585	-	-	-	11,455
Slovakia	-	138	-	-	-	-	-	138
Spain	974	-	-	2,149	-	-	-	3123
United Kingdom	19,242	-	-	-	-	-	-	19,242
Argentina	-	49,867	-	-	-	-	-	49,867
Indonesia	-	-	34,082	-	-	-	-	34,082
Brazil	-	17,460	-	-	42	-	-	17,502
Canada	8876	1842	-	-	544	921	-	12,184
Ukraine	10,551	586	-	-	-	-	-	11,137
US	293	9253	-	-	502	209	-	10,258
Malaysia	-	-	8876	-	-	-	-	8876
Paraguay	126	7746	-	-	-	-	-	7872
Russia	3350	1884	-	-	-	-	-	5234
China	-	42	-	-	-	2805	-	2847
Others	4145	586	544	-	-	42	-	5317
Total	198,924	92,910	43,712	18,590	7746	53,468	126	**415,476**
Share	48%	22%	11%	4%	2%	13%	0%	

Table 2. Origin of crops underlying the European bioethanol consumption mix in 2010 (10^3 GJ) [2,14]; gray color indicates exclusion from water footprint calculation as the origin is unknown.

	Wheat	Maize	Barley	Rye	Triti-cale	Sugar Beet	Wine	Sugar Cane	Others	Total
European Union	24,326	14,403	2428	3391	837	30,691	4229	-	1382	81,688
Austria	-	382	-	-	-	1119	-	-	-	1501
Belgium	-	-	-	-	-	1430	-	-	-	1430
Bulgaria	1072	536	-	-	-	-	-	-	-	1609
Croatia	-	437	-	-	-	-	-	-	-	437
Czech Republic	989	-	72	-	-	1209	-	-	-	2270
Denmark	871	-	179	199	-	-	-	-	-	1249
Finland	-	-	86	-	-	-	-	-	-	86
France	8124	3511	467	-	-	10,852	-	-	-	22,953
Germany	5264	1023	468	1775	-	7370	-	-	-	15,900
Greece	334	510	-	-	-	-	-	-	-	843
Hungary	1072	1568	-	-	-	-	-	-	-	2640
Ireland	-	-	75	-	-	-	-	-	-	75
Italy	1475	1517	-	-	-	-	-	-	-	2991
The Netherlands	-	-	-	-	-	1849	-	-	-	1849
Poland	1992	943	132	1272	-	3419	-	-	-	7758
Portugal	-	198	-	-	-	-	-	-	-	198
Romania	1535	2646	70	-	-	-	-	-	-	4251
Spain	1599	1132	455	145	-	860	-	-	-	4190
Sweden	-	-	88	-	-	-	-	-	-	88
United Kingdom	-	-	321	-	-	2583	-	-	-	2904
Brazil	-	335	-	-	-	-	-	9798	-	10,133
USA	84	5108	-	-	-	-	-	-	-	5192
Peru	-	-	-	-	-	-	-	1089	-	1089
Switzerland	1047	-	-	-	-	-	-	-	-	1047
Bolivia	-	-	-	-	-	-	-	837	-	837
Ukraine	209	293	-	-	-	84	-	-	-	586
Egypt	-	-	-	-	-	-	-	628	-	628
Guatemala	-	-	-	-	-	-	-	586	-	586
Argentina	-	84	-	-	-	-	-	209	-	293
Cuba	-	-	-	-	-	-	-	251	-	251
Other	419	293	-	-	-	-	-	670	84	1465
Total	26,085	20,516	2412	3391	837	30,774	4229	14,068	1465	**103,780**

Table 3. Specific irrigation water consumption of crops (1996–2005) underlying the European biodiesel consumption (m^3/GJ) [11].

	Rapeseed	Soybean	Oil Palm	Sunflower
European Union				
Austria	-	0.00	0.00	-
Bulgaria	-	-	-	0.38
Croatia	-	0.00	-	0.00
Czech Republic	0.00	-	-	-
Denmark	0.00	-	-	-
France	0.43	71.54	-	1.55
Germany	0.00	-	0.00	-
Greece	-	-	-	90.94
Hungary	3.87	0.98	-	0.21
Italy	-	20.95	-	9.60
Lithuania	0.00	-	-	-
Poland	0.00	-	-	-
Romania	0.00	103.10	-	16.44
Slovakia	-	24.19	-	-
Spain	6.02	-	-	128.06
United Kingdom	0.00	-	-	-
Argentina	-	0.85	-	-
Indonesia	-	-	0.00	-
Brazil	-	0.13	-	-
Canada	0.00	0.00	-	-
Ukraine	0.00	0.00	-	-
US	1.76	14.79	-	-
Malaysia	-	-	0.00	-
Paraguay	0.00	0.00	-	-
Russia	0.00	0.00	-	-
China	-	39.89	-	-

Table 4. Specific irrigation water consumption of crops (1996–2005) underlying the European bioethanol consumption (m^3/GJ) [11].

	Wheat	Maize	Barley	Rye	Sugar beet	Sugar cane
European Union						
Austria	-	0.00	-	-	1.40	-
Belgium	-	-	-	-	0.01	-
Bulgaria	0.00	1.65	-	-	-	-
Croatia	-	0.03	-	-	-	-
Czech Republic	0.00	-	0.00	-	0.00	-
Denmark	0.62	-	0.02	0.00	-	-
Finland	-	-	0.00	-	-	-
France	0.13	9.00	0.40	-	0.48	-
Germany	0.00	0.21	0.00	0.00	1.04	-
Greece	2.83	49.38	-	-	-	-
Hungary	0.18	0.11	-	-	-	-
Ireland	-	-	0.00	-	-	-
Italy	1.58	10.75	-	-	-	-
The Netherlands	-	-	-	-	0.08	-
Poland	0.04	0.37	0.00	0.02	0.22	-
Portugal	-	86.00	-	-	-	-
Romania	4.47	1.93	0.00	-	-	-
Spain	4.68	40.43	8.23	0.00	21.63	-
Sweden	-	-	0.00	-	-	-
United Kingdom	-	-	0.00	-	0.33	-
Brazil	-	0.05	-	-	-	2.32
USA	9.04	6.31	-	-	-	-
Peru	-	-	-	-	-	26.09
Switzerland	0.00	-	-	-	-	-
Bolivia	-	-	-	-	-	3.92
Ukraine	1.85	8.66	-	-	1.02	-
Egypt	-	-	-	-	-	61.93
Guatemala	-	-	-	-	-	5.40
Argentina	-	1.40	-	-	-	11.70
Cuba	-	-	-	-	-	21.53

2.2. Impact Assessment

After the water inventory was determined, an impact assessment has been conducted in order to evaluate the local consequences resulting from the water consumption in the different regions.

For this, the water accounting and vulnerability evaluation (WAVE) model [19] and the impact assessment model of Pfister and colleagues [20] have been applied. Both methods rely on local freshwater scarcity which is calculated by relating annual water use (in case of WAVE water consumption) to annual freshwater availability (groundwater recharge and surface run-off). Such ratios have been determined for more than 11,000 river basins on a global level by means of the hydrological model WaterGAP2 [21] which provides annual average hydrological information based on the climate normal period (1961–1990). Based on these scarcity ratios and further hydrologic parameters, the WAVE model determines regional water depletion indexes (WDI) which denote the risk that water consumption leads to freshwater depletion (Figure 1).

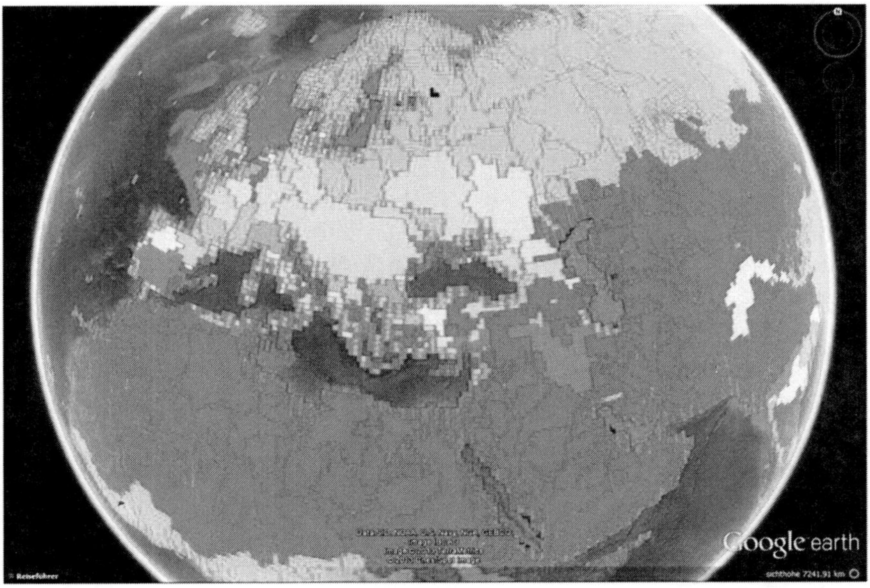

Figure 1. Water depletion index (WDI), which denotes the risk that water consumption leads to freshwater depletion, ranging from 0.01 (blue) to 1.00 (red)—Google Earth layer can be downloaded free of charge at http://www.see.tu-berlin.de/WAVE.

The model of Pfister and colleagues also provides a general scarcity based indicator (water stress index) and additionally assesses damages resulting from water consumption specifically for human health, ecosystems, and resources according to the procedure described in ref. [20]. While all impact factors have been developed on the spatial resolution of river basins, national average factors have been derived by creating water consumption weighted averages for each country.

The overall impacts per GJ of biodiesel, bioethanol, fossil diesel, and fossil gasoline have been determined by multiplying their spatially explicit water inventories by the corresponding national impact factors and aggregating the local impacts.

2.3. Influence of Biofuels on Carbon and Water Footprint of Cars along the Life Cycle

In order to compare the carbon and water footprint of cars run by fossil fuels and biofuels, GHG emissions and water consumption of a car's production and recycling phases have been estimated based on existing studies [17,22,23]. Assuming a mileage of 200,000 km and an average fuel consumption of 9 L gasoline per 100 km (and 13.8 L bioethanol, both equals 369 MJ) the global warming potential and the water consumption of a car's use phase has been estimated. An average emission of 2.5 kg CO_2 per L gasoline and a reduction of 60% [24] have been assumed. For water, the consumption figures determined for gasoline and biofuels in this paper have been used. In addition to the volumetric information, impacts resulting from water consumption along the product life cycle have been assessed by means of the WAVE model [19].

3. RESULTS AND DISCUSSION

3.1. Water Inventories

By connecting the information regarding the crops and their origins underlying European biodiesel and bioethanol consumption (Table 1 and Table 2) with the crop-specific national irrigation requirements (Table 3 and Table 4), the total volume of water consumed around the globe for the provision of European biofuels has been determined (Table 5 and Table 6).

Table 5. Total (blue) water consumption of crops underlying the European biodiesel consumption (10^6 m^3).

	Rapeseed	Soybean	Oil palm	Sunflower	Total	Share
European Union	41.3	119.1	0.0	426.2	586.7	76%
Austria	-	0.0	-	-	0.0	0%
Bulgaria	-	-	-	1.5	1.5	0%
Croatia	-	0.0	-	0.0	0.0	0%
Czech Republic	0.0	-	-	-	0.0	0%
Denmark	0.0	-	-	-	0.0	0%
France	17.0	27.4	-	5.1	49.5	6%
Germany	0.0	-	-	-	0.0	0%
Greece	-	-	-	63.9	63.9	8%
Hungary	18.5	0.3	-	0.6	19.4	3%
Italy	-	34.4	-	4.5	38.9	5%
Lithuania	0.0	-	-	-	0.0	0%
Poland	0.0	-	-	-	0.0	0%
Romania	0.0	53.7	-	75.4	129.1	17%
Slovakia	-	3.3	-	-	3.3	0%
Spain	5.9	-	-	275.2	281.0	36%
United Kingdom	0.0	-	-	-	0.0	0%
Argentina	-	42.4	-	-	42.4	6%
Indonesia	-	-	0.0	-	0.0	0%
Brazil	-	2.3	-	-	2.3	0%
Canada	0.0	0.0	-	-	0.0	0%
Ukraine	0.0	0.0	-	-	0.0	0%
US	0.5	136.9	-	-	137.4	18%
Malaysia	-	-	0.0	-	0.0	0%
Paraguay	0.0	0.0	-	-	0.0	0%
Russia	0.0	0.0	-	-	0.0	0%
China	-	2	-	-	1.7	0%
Total	41.8	302.3	0.0	426.2	**770.4**	100%
Share	5%	39%	0%	55%	100%	

Table 6. Total (blue) water consumption of crops underlying the European bioethanol consumption (10^6 m^3).

	Wheat	Maize	Barley	Rye	Sugar beet	Sugar cane	Total	Share
European Union	19.5	142.6	3.9	0.0	34.8	0.0	200.9	59%
Austria	-	0.0	-	-	1.6	-	1.6	0%
Belgium	-	-	-	-	0.0	-	0.0	0%
Bulgaria	0.0	0.9	-	-	-	-	0.9	0%
Croatia	-	0.0	-	-	-	-	0.0	0%
Czech Republic	0.0	-	0.0	-	0.0	-	0.0	0%
Denmark	0.5	-	0.0	0.0	-	-	0.5	0%
Finland	-	-	0.0	-	-	-	0.0	0%
France	1.1	31.6	0.2	-	5.2	-	38.0	11%
Germany	0.0	0.2	0.0	0.0	7.7	-	7.9	2%
Greece	0.9	25.2	-	-	-	-	26.1	8%
Hungary	0.2	0.2	-	-	-	-	0.4	0%
Ireland	-	-	0.0	-	-	-	0.0	0%
Italy	2.3	16.3	-	-	-	-	18.6	5%
The Netherlands	-	-	-	-	0.1	-	0.1	0%
Poland	0.1	0.3	0.0	0.0	0.8	-	1.2	0%
Portugal	-	17.0	-	-	-	-	17.0	5%
Romania	6.9	5.1	0.0	-	-	-	12.0	4%
Spain	7.5	45.8	3.7	0.0	18.6	-	75.6	22%
Sweden	-	-	0.0	-	-	-	0.0	0%
United Kingdom	-	-	0.0	-	0.9	-	0.9	0%
Brazil	-	0.0	-	-	-	22.7	22.7	7%
USA	0.8	32.2	-	-	-	-	33.0	10%
Peru	-	-	-	-	-	28.4	28.4	8%
Switzerland	0.0	-	-	-	-	-	0.0	0%
Bolivia	-	-	-	-	-	3.3	3.3	1%
Ukraine	0.4	2.5	-	-	0.1	-	3.0	1%
Egypt	-	-	-	-	-	38.9	38.9	11%
Guatemala	-	-	-	-	-	3.2	3.2	1%
Argentina	-	0.1	-	-	-	2.4	2.6	1%
Cuba	-	-	-	-	-	5.4	5.4	2%
Total	21	178	4	0	35	104	**341.3**	100%
Salıre	6%	52%	1%	0%	10%	31%	100%	

Dividing the total water consumption by the total biofuel consumption, an average water consumption of 1.9 m^3/GJ for biodiesel and 3.3 m^3/GJ for bioethanol has been determined. It should be noted that in both cases, the specific water consumption can be very diverse depending on the underlying crop and country. While no irrigation is needed to cultivate crops for biodiesel in the UK, Poland, and Germany, on average 90 m^3 of irrigation water are consumed to produce 1 GJ of biodiesel in Spain or from Spanish crops. Bioethanol or underlying crops can be produced without irrigation in Czech Republic and Switzerland. In contrast, the production of bioethanol or underlying crops in Portugal consumes 86 m^3/GJ.

The analysis of water consumption for fossil fuels resulted in water consumption figures of 0.032 m^3/GJ for diesel and 0.081 m^3/GJ for gasoline. Hence, biodiesel causes about 70 times and bioethanol 44 times the water consumption compared to their fossil alternatives (Figure 2). Taking into account these significant differences, it is assumed that data availability related differences in the reference years (origin of feedstock 2010, water consumption of biofuels 1996–2005, and fossil fuels 2013) do not change the general finding from this study.

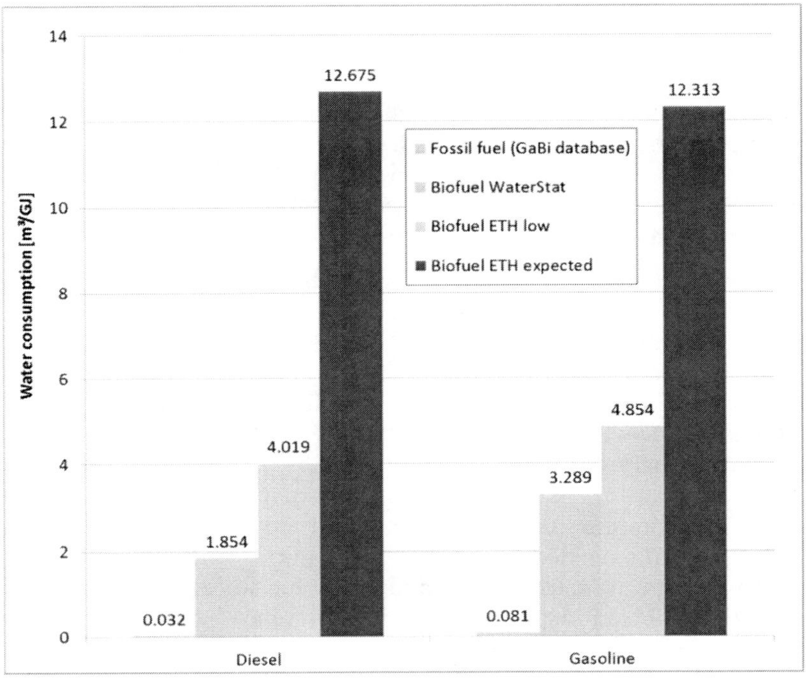

Figure 2. Water consumption per unit of energy derived from average fossil diesel, biodiesel, fossil gasoline, and bioethanol consumed in Europe. The figure includes the estimate form the WaterStat database [11], and the ETH database low [25] and normal estimate [26] included as a sensitivity check.

3.2. Uncertainties in Volumetric Results

Since irrigation water consumption data is based on global models, uncertainties in the estimates should be analyzed. Therefore, additional irrigation water consumption data from the ETH databases [25,26] has been applied which provides an expected and a low estimate for the blue water footprint of biofuels. As shown in Tables S1–S4 in the supporting information, the water consumption estimates per biofuels or specific origin can vary considerably, depending on the model used: Spain is getting a lower share in the blue water footprint of EU's biofuel consumption, while France, Romania and Argentina get a higher percentage. This is based on differences in the underlying ETH model which shows relatively high differences in water consumption for some regions compared to WaterStat. While the ETH data is based on spatially explicitly modeled yields, WaterStat calculates the yields as a function of water availability (and therefore a function of modeled irrigation). ETH data corrects calculated theoretical irrigation water demand based on a spatially explicit irrigation dataset, while WaterStat calculates water availability of precipitation and irrigation in soil moisture as a function of a different irrigation model. Since both approaches have advantages and disadvantages and cannot reflect the real water consumption, the observed difference can be expected, since all the input data required for the models are uncertain, especially since they are available on high spatial resolution (~10 km) on global coverage.

However, total global agricultural water consumption of the WaterStat database is between the ETH expected and low estimates, so there is no general trend between the two databases—which appears to be the case according to the results shown in this paper. Based on the analysis in this paper, the average water consumption for EU results to 4.0–12.7 m^3/GJ for biodiesel and 4.9–12.3 m^3/GJ for bioethanol, which is considerably higher than the numbers based on WaterStat (Figure 2). This variability among different data also indicates that the difference of EU bioethanol and biodiesel consumption is not significant and more detailed assessment of specific origins is needed to compare two specific options. Uncertainty is even more relevant since the variability of blue water consumption for a crop within one country can be very high as reported for the ETH model [21]: for soybean production, the coefficient of variation (CV) is 2.9 for Argentina and 2.7 for the US, while for Spain, the CV of sunflower and maize production is 0.9.

However, the results based on the two databases agree that the blue water footprint of biofuels is ~2 orders of magnitude higher compared to their fossil alternatives.

The average water consumption caused by European biofuel consumption is significantly lower than the world average water consumption of biofuels. Based on ETH data, the global average biofuel production has roughly 2–2.5 times the water consumption of EU consumption (Table S5). Based on WaterStat data, the global average is around two orders of magnitude higher for biodiesel (217 m^3/GJ for soybean to 335 m^3/GJ for jatropha) and one order of magnitude for bioethanol (18 m^3/GJ for cassava to 182 m^3/GJ for sorghum) [8]. This can be explained by the fact that 60% and 79% of the crops underlying European biodiesel and bioethanol consumption are grown within the European Union where, except for Southern European countries, irrigation demand is relatively low. Moreover, as it can be seen in Table 3 and Table 4, also the imported crops are derived from countries with mainly low irrigation needs, such as Soybean from Argentina or sugar cane from Brazil. However, the enormous difference between European and global average water consumption might also be a result of the high model uncertainty as discussed above.

Since absolute volumes do not allow for an assessment of local impacts, a regionalized water inventory has been established for both fossil and biofuels according to the methodology described in the previous chapter. The following maps show the relative local shares of water consumed in the production of average biodiesel and fossil diesel (Figure 3a) and bioethanol and fossil gasoline (Figure 3b). Taking into account the relative presentation, the maps allow for a comparison of local water consumption per individual fuel only. Hence, the bars of different fuels cannot be compared to each other since they are based on different absolute water consumption figures per GJ of fuel.

As it can be seen from the maps, water consumption in biodiesel and diesel production takes place in 12 and 16 countries, respectively. The production of bioethanol and gasoline consumes water in 23 and 14 countries. For biodiesel, 72% of the water consumption occurs in Spain (sunflower seed), the US (soybean), and Romania (sunflower seed and soybean). For bioethanol, water consumption is distributed more homogeneously. Nevertheless, Spain (22%, mainly maize and sugar beet), Egypt (11%, sugar cane), France (11%, mainly maize), and the US (10%, mainly maize) can be regarded as the main contributors. The relative spatial distribution for biofuel based on the alternative database is shown per country in the supporting information (Figure S1) indicating differences in relative importance.

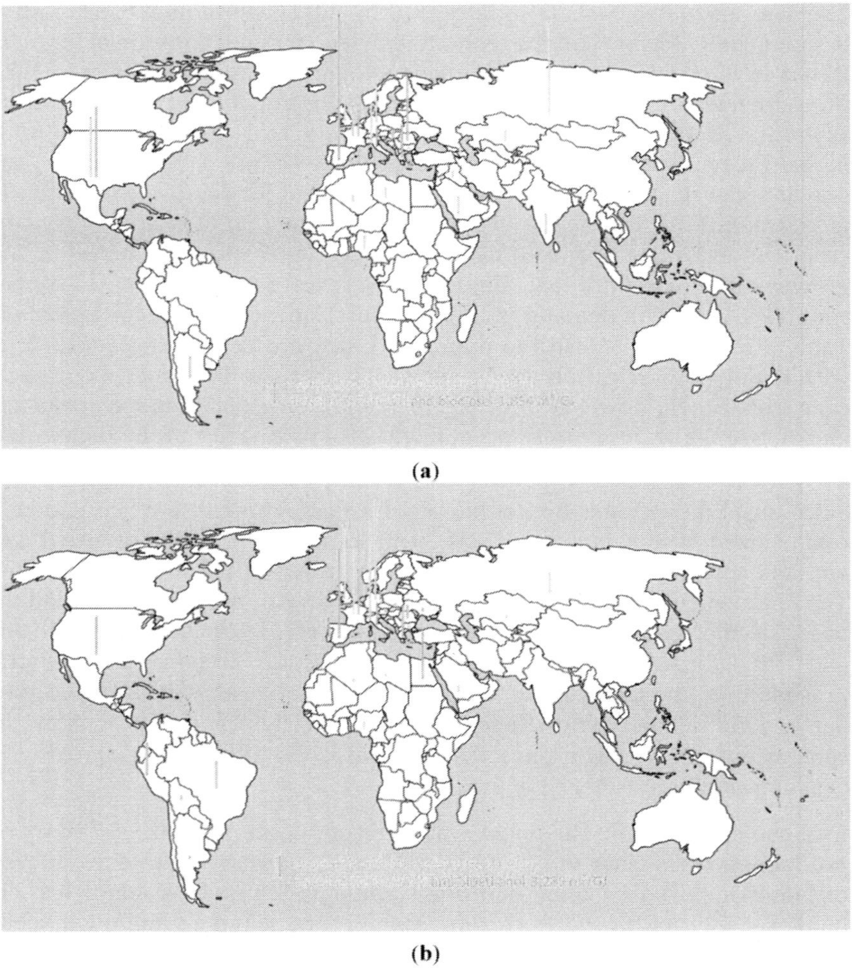

Figure 3. Relative local water consumption occurring in the production of (**a**) fossil diesel (blue bars) and biodiesel (green bars) and (**b**) gasoline (blue bars) and bioethanol (green bars) consumed in the European Union.

3.3. Impact Assessment

By multiplying the spatially explicit water inventories shown in Figure 3 with the corresponding regional characterization factors provided by the WAVE model [19] and the model of Pfister and colleagues [20], impacts from water consumption in these countries have been assessed (Figure 3).

It should be noted that for biofuels only the consumption figures determined based on WaterStat are considered in this impact assessment. Since only water consumption but no water pollution is considered in this work, this study represents a water availability footprint according to ISO 14046 [9].

Comparing biodiesel and fossil diesel, it has been shown that the differences on the inventory level (varying by a factor of 60) are in a similar range on the impact assessment level (factor 55–74). The only exception from this trend is the impact category damage to human health. Here, biodiesel causes similar impacts as fossil diesel despite the significantly higher water consumption. This can be explained by the locations of water consumption in biodiesel production. As shown in Figure 3a, water consumption occurs mainly in Europe and in the US. Even though physical water scarcity is relevant in many of those countries, their high degree of development avoids health damages resulting from water stress.

Concerning bioethanol, Figure 4 shows that differences on the impact assessment level are significantly higher (factor 64–246) than on the inventory level (factor 40). The reason for this can be found in the origin of the feedstock (Figure 3b) and the local water scarcity which leads to higher differences on the impact than on the volumetric level compared to gasoline. For instance, crops grown for European bioethanol consumption cause irrigation water consumption in water stressed countries like Egypt or Spain, while gasoline production has main water consumption in North-Western Europe (Figure 3b) with low water scarcity and no human health impacts.

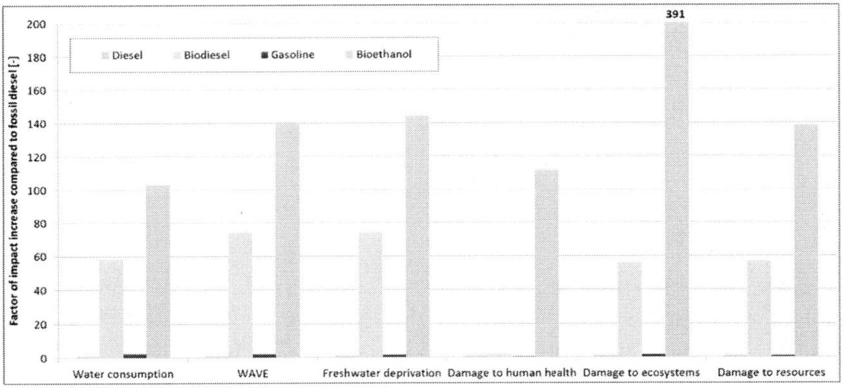

Figure 4. Water consumption and resulting local impacts presented in relation to fossil diesel, i.e., factor of increase compared to fossil diesel in each category.

In order to identify the most relevant crops and countries, the origin of the impacts resulting from water consumption have been analyzed in more detail by means of the impact category freshwater depletion from the WAVE model [19]. As it can be seen from Figure 5a, the irrigation of mainly sunflower seed in Spain causes 50% of the impacts, even though it constitutes less than 0.9% of the feedstock in European biodiesel production. In case of bioethanol production, the irrigation of sugar cane in Egypt, which constitutes only 0.7% of European bioethanol production, causes 20% of the impacts. For agricultural products, where blue water consumption is largely depending on climate this effect is generally observed: places with high irrigation typically show higher water scarcity. However, it should be noted that average country factors have been used for both water consumption figures and characterization factors. Especially in large countries like the US or China, an evaluation on the level of river basins is more adequate. Moreover, considering the uncertainty of blue water estimates discussed above, the distribution of impacts should be considered as indicative rather than absolute shares.

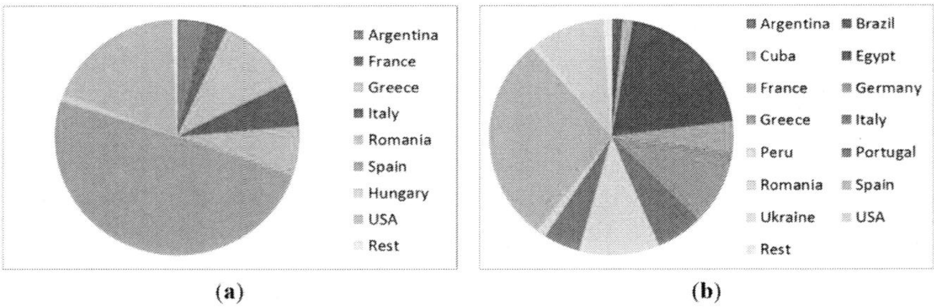

Figure 5. Relative contribution of countries to the results of the impact category freshwater depletion from the WAVE model [19] for (**a**) biodiesel and (**b**) bioethanol.

3.4. Influence of Biofuels on Carbon and Water Footprint of Cars along the Life Cycle

So far, water consumption and related impacts have been analyzed for fossil fuels and biofuels solely. Yet, taking a holistic perspective, the influence of biofuels on the water footprint of a car along its entire lifecycle is more relevant. In order to address this aspect and to compare a car's carbon and water footprint, the global warming potential (Figure 6a), the volumetric water consumption (Figure 6b) and the resulting impacts (Figure 6c) of a car run by gasoline and bioethanol has been determined throughout the life cycle as described in Section 2.2. It should be noted that fuel production is assigned to the use phase of a car.

As it can be seen in Figure 5a, bioethanol can reduce the carbon footprint of a car by about 50% compared to a car run by fossil gasoline. However, taking into account the 40 times higher water consumption of bioethanol compared to fossil gasoline, the life cycle based water consumption increases from about 105 to 1979 m³ (Figure 6b). On the impact assessment level (Figure 6c), the increase in freshwater depletion is even more significant which can be explained by the local water scarcity in the countries providing the underlying crops.

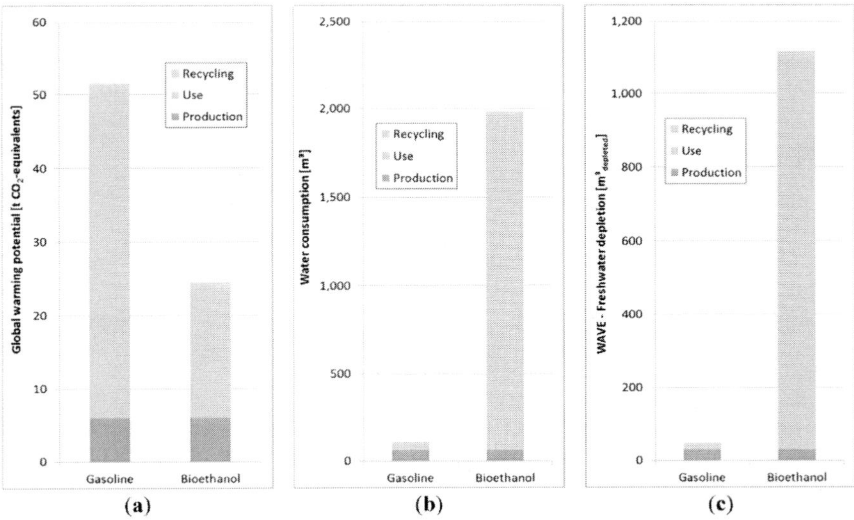

Figure 6. (**a**) Global warming potential (**b**) Volumetric water consumption (**c**) Impacts of water consumption (WAVE–freshwater depletion) along the life cycle of an average passenger car run by gasoline and bioethanol.

Even potential uncertainties in the water consumption figures of the car's production, fuel production, and recycling phase are not expected to affect the main finding of this work: European biofuels cause significantly higher water consumption in car transportation than their fossil alternatives. Thus, a reduction in a car's carbon footprint achieved by the use of biofuels is accompanied by a significant increase in its water footprint.

4. CONCLUSIONS

While biofuels are promoted to decrease the carbon footprint of car transportation, little information regarding the global water footprint of biofuels consumed in Europe is available. This study compares the water consumption and associated impacts resulting from the production of 1 GJ fuel of average European biodiesel, diesel, bioethanol, and gasoline.

The ultimate origin of feedstock underlying European biodiesel and bioethanol consumption has been combined with the irrigation requirements of different crops in different countries. As a result, a (blue) water consumption of 1.9 m^3 per GJ of European biodiesel and 3.3 m^3 per GJ of bioethanol has been determined. Even though this represents an increase by a factor of 60 and 40 compared to fossil diesel and gasoline, these figures are low compared to global average water consumption data for biofuels. However, the estimates are quite uncertain as a comparison with an alternative database resulted in higher water consumption (by a factor 1.5 to 6.8).

In order to assess potential local impacts resulting from this water consumption regional information on where the water consumption occurs is required. This analysis revealed that average European biodiesel causes water consumption in 12 countries and bioethanol is responsible for water consumption in 23 countries. In contrast, fossil diesel and gasoline are responsible for water consumption in 16 and 14 countries, respectively.

These regional water consumption figures have been multiplied by local characterization factors derived from the WAVE model [19] and the impact assessment method developed by Pfister and colleagues [20]. While the difference between biodiesel and fossil diesel is similar on the inventory and the impact assessment levels (factor 55–74, except human health impacts), bioethanol shows larger differences compared to gasoline on the impact level (factor 64–246) than on the volumetric level (factor 40). The reason for the significantly higher impacts of biofuels compared to fossil fuels can be found in the fact that relatively large shares of water consumption in the production of biofuels are caused in relatively dry countries. For instance, the irrigation of mainly sunflower seed in Spain causes 52% of the impacts, even though it constitutes only 0.5% of the feedstock in European biodiesel production. In case of bioethanol production, the irrigation of sugar cane in Egypt, which constitutes only 0.4% of European bioethanol production, causes 20% of the impacts. Yet, it should be noted that these conclusions are attenuated when considering the ETH databases.

By means of a case study on passenger cars, the trade-off between water and carbon footprints has been illustrated. While biofuels generally cause less greenhouse gas emissions than fossil fuels (factor ~0.5), the water consumption (factor ~19) and especially the resulting local impacts (factor ~24) are more severe.

ACKNOWLEDGMENTS

The authors wish to express sincere thanks to Ivetta Gerasimchuk (Chatham House) who provided insight into the data used to determine the origin of feedstock used for biofuel production. Moreover, the valuable feedback of Gerhard Brankatschk (OVID) concerning crop cultivation and biofuel production is highly appreciated.

AUTHOR CONTRIBUTIONS

Markus Berger accomplished the survey and wrote a draft of the manuscript. Stephan Pfister added an analysis and discussion by using alternative ETH databases for the irrigation water consumption needed to produce the crops used for biodiesel and bioethanol production. Vanessa Bach developed the software tool which creates the spatially explicit water inventories used as a base for the impact assessment of this work. Matthias Finkbeiner reviewed the manuscript and provided valuable comments to deepen the discussion and conclusion.

CONFLICTS OF INTEREST

The authors declare no conflict of interest.

REFERENCES

1. Directive 2009/28/EC. Directive 2009/28/ec of the european parliament and of the council of 23 april 2009 on the promotion of the use of energy from renewable sources. Available online:

http://eur-lex.europa.eu/legal-content/en/TXT/?uri=CELEX:32009L0028 (accessed on 6 February 2015).

2. Gerasimchuk, I. Energy, Environment and Resources eer pp 2012/04: Biofuel Policies and Feedstock in the EU; Chatham House: London, UK, 2013.

3. ISO. Environmental Management—Life Cycle Assessment–Requirements and Guidelines (ISO 14044:2006); International Organisation for Standardisation, Ed.; ISO: Geneva, Switzerland, 2006.

4. ISO. ISO 14064-1 Greenhouse Gases—Part 1—Specification for the Quantification, Monitoring and Reporting of entity Emissions and Removals; ISO 14064-2 Greenhouse Gases—Part 2—Specification for the Quantification, Monitoring and Reporting of Project Emissions and Removals; ISO 14064-3 Greenhouse Gases—Part 3—Specification and Guidance for Validation, Verification and Certification; International Organization for Standardization, Ed.; ISO: Geneva, Switzerland, 2006.

5. Zah, R.; Böni, H.; Gauch, M.; Hischier, R.; Lehmann, M.; Wäger, P. Ökobilanz von Energieprodukten: Ökologische Bewertung von Biotreibstoffen; Empa, Technology and Society Lab: St. Gallen, Switzerland, 2007.

6. Sanchez, S.T.; Woods, J.; Akhurst, M.; Brander, M.; O'Hare, M.; Dawson, T.P.; Edwards, R.; Liska, A.J.; Malpas, R. Accounting for indirect land-use change in the life cycle assessment of biofuel supply chains. J. R. Soc. Interface **2012**, 9, 1105–1119.

7. Finkbeiner, M. Indirect land use change-help beyond the hype? Biomass Bioenergy **2014**, 62, 218–221.

8. Gerbens-Leenes, W.; Hoekstra, A.Y.; van der Meer, T.H. The water footprint of bioenergy. Proc. Natl. Acad. Sci. USA **2009**, 106, 10219–10223.

9. ISO. Water Footprint-Principles, Requirements and Guidance; International Organization for Standardization, Ed.; ISO: Geneva, Switzerland, 2014.

10. Gerbens-Leenes, W.; Hoekstra, A.Y. The water footprint of biofuel-based transport. Energy Environ. Sci. **2011**, 4, 2658–2668.

11. Mekonnen, M.M.; Hoekstra, A.Y. The green, blue and grey water footprint of crops and derived crop products. Hydrol. Earth Syst. Sci. **2011**, 15, 1577–1600.

12. Bayart, J.B.; Bulle, C.; Koehler, A.; Margni, M.; Pfister, S.; Vince, F.; Deschenes, L. A framework for assessing off-stream freshwater use in lca. Int. J. Life Cycle Assess. **2010**, 15, 439–453.

13. Ecofys. Renewable energy progress and biofuels sustainability. Available online: http://www.ecofys.com/en/publication/progress-in-renewable-energy-and-biofuels-sustainability/ (accessed on 7 May 2015).

14. FAOSTAT. Food and aggriculture organization of the united nations statistical devision. Available online: http://faostat3.Fao.Org/home/e (accessed on 15 January 2015).

15. PE International. Gabi LCI database. Available online: http://www.gabi-software.com (accessed on 27 March 2015).

16. Berger, M.; Finkbeiner, M. Methodological challenges in volumetric and impact oriented water footprints. J. Ind. Ecol. **2013**, 17, 79–89.

17. Berger, M.; Warsen, J.; Krinke, S.; Bach, V.; Finkbeiner, M. Water footprint of european cars: Potential impacts of water consumption along automobile life cycles. Environ. Sci. Technol. **2012**, 46, 4091–4099.

18. Martinez-Blanco, J.; Berger, M.; Bach, V.; Finkbeiner, M. Promoting the application of water footprints: A tool for the regionalization of generic water inventories. In Proceedings of the Life Cycle Assessment XIII, Orlando, FL, USA, 1–3 October 2013.

19. Berger, M.; van der Ent, R.; Eisner, S.; Bach, V.; Finkbeiner, M. Water accounting and vulnerability evaluation (wave)—Considering atmospheric evaporation recycling and the risk of freshwater depletion in water footprinting. Environ. Sci. Technol. **2014**, 48, 4521–4528.

20. Pfister, S.; Koehler, A.; Hellweg, S. Assessing the environmental impacts of freshwater consumption in lca. Environ. Sci. Technol. **2009**, 43, 4098–4104.

21. Flörke, M.; Kynast, E.; Bärlund, I.; Eisner, S.; Wimmer, F.; Alcamo, J. Domestic and industrial water uses of the past 60 years as a mirror of socio-economic development: A global simulation study. Glob. Environ. Change **2013**, 23, 144–156.

22. Kim, H.C.; Wallington, T.J.; Mueller, S.A.; Berdish, D.; Bras, B.; Tejada, F.; Guldberg, T. Water use in the life cycle of light-duty vehicles. In Proceedings of the 10th International Conference on EcoBalance, Tokyo, Japan, 21–23 November 2012.

23. Daimler. Mercedes-benz models with environmental certificates. Available online: http://www.Daimler.Com/sustainability /product-responsibility (accessed on 18 April 2015).

24. Van den Bos, A.; Hamelinck, C. Greenhouse Gas Impact of Marginal Fossil Fuel Use; Ecofys: Utrecht, The Netherlands, 2014.

25. Pfister, S.; Bayer, P.; Koehler, A.; Hellweg, S. Environmental impacts of water use in global crop production: Hotspots and trade-offs with land use. Environ. Sci. Technol. **2011**, 45, 5761–5768.

26. Pfister, S.; Bayer, P. Monthly water stress: Spatially and temporally explicit consumptive water footprint of global crop production. J. Clean. Prod. **2014**, 73, 52–62.

9

CHAPTER

EFFECTS OF ENVIRONMENTAL FACTORS AND NUTRIENT AVAILABILITY ON THE BIOCHEMICAL COMPOSITION OF ALGAE FOR BIOFUELS PRODUCTION: A REVIEW

Ankita Juneja [1], Ruben Michael Ceballos [2] and Ganti S. Murthy [1,*]

[1] *Biological and Ecological Engineering, Oregon State University, Corvallis, OR 97331, USA*

[2] *Native American Research Laboratory, Division of Science and Mathematics, University of Minnesota-Morris, Morris, MN 56267, USA*

ABSTRACT

Due to significant lipid and carbohydrate production as well as other useful properties such as high production of useful biomolecular substrates (e.g., lipids) and the ability to grow using non-potable water sources, algae are being explored as a potential high-yield feedstock for biofuels production. In both natural and engineered systems, algae can be exposed to a variety of environmental conditions that affect growth rate and cellular composition. With respect to the latter, the amount of carbon fixed in lipids and carbohydrates (e.g., starch) is highly influenced by environmental factors and nutrient availability. Understanding synergistic interactions between multiple environmental variables and nutritional factors is required to develop sustainable high productivity bioalgae systems, which are essential for commercial biofuel production. This article reviews the effects of environmental factors (i.e., temperature, light and pH) and nutrient availability (e.g., carbon, nitrogen, phosphorus, potassium, and trace metals) as well as cross-interactions on the biochemical composition of algae with a special focus on carbon fixation and partitioning of carbon from a biofuels perspective.

Keywords: algae; biochemical composition; environmental effect; biofuel production

1. INTRODUCTION

Increasing demand for energy and global warming are two major challenges facing modern society. Dependence on fossil fuels for meeting increasing energy demands is unsustainable due to increasing levels of consumption and a dearth in discovery of new sources for these non-renewables. This concern has motivated researchers to focus on the development of alternative energy sources including solar, wind, water, and biomass. Biofuels are alternatives to liquid fossil fuels and are produced from sugar, starch, cellulosic or lipid-rich substrates. These substrates can be derived from feedstocks such as cereal crops, including corn and wheat [1]; sugar crops, including sorghum and sugarcane [2]; energy crops, such as switchgrass [3,4]; agricultural wastes, including straws and corn stover [5,6,7,8]; municipal wastes; and, several aquatic species. Currently, ethanol is produced from corn and sugarcane in significant volumes as a supplemental fuel and as a partial substitute for gasoline. Production and use of ethanol as a transportation fuel results in a net reduction of greenhouse gas (GHG) emissions. This claim is based on the idea that carbon dioxide emissions produced during the processing of biomass and from the use of biofuels is readily sequestered.

Biofuels produced from cereal crops and sugarcane are commonly called first-generation biofuels (FGB). The large-scale production (and use) of FGB products as competitive substitutes for fossil fuels is hindered by several limitations including: intensive agricultural inputs, land requirements, and trade-offs between food crop and fuel crop production (i.e., the food vs. fuel debate) [9]. Therefore, although they are renewable, FGBs alone are not a viable solution for solving global liquid fuel demands. Second-generation biofuels (SGB) are fuels derived from lignocellulosic biomass. Production of SGBs circumvents several of the negative outcomes associated with FGBs [10]. At present, the development of competitive SGB products is at various stages of research and pilot demonstrations. A few products have been commercialized. However, SGB feedstock production still requires agricultural inputs, land, and freshwater that could be used for food crops. Thus, SGBs are also subject to the food-versus-fuel dilemma in the long term. Producing liquid fuels from aquatic organisms, such as algae, is considered to be the third-generation in biofuels (TGB). The use of TGB feedstocks, which contain significant amounts of lipids and carbohydrates, from which biodiesel and bioethanol products may be produced, avoids most of the limitations noted for FGBs and SGBs. Arguably, this includes overcoming the food vs. fuel dilemma. Most aquatic plant feedstocks are championed as a viable source of lipids for the production of bio-oil [11]. Specifically, via thermochemical conversions or biochemical conversions, algae can be used to produce: biofuel oil and gas; or, bioethanol, biodiesel, and biohydrogen, respectively [12].

A recent resurgence of interest in algal-based TGBs is attributed to significant benefits associated with TGB production. These benefits include: year-round production; higher productivity compared to terrestrial crops; the potential for off-shore production (and, thus, non-competition with food crops); reduced need for arable land; reduced need for water treatment; and potential advantages in nutrient cycling [13,14,15].

Earlier research demonstrated that under select conditions, algae have the potential to produce 40 times the amount of oil for biodiesel production compared to oilseed crops (i.e., soy and canola) per unit land area [16]. However, such advantages over FGBs and SGBs have only been established in academic/research facilities. Presently, in the absence of publicly available data, it is unknown whether such gains can be realized on a commercial scale. Therefore, the economic potential of algal-based biofuels to significantly impact current and future fuels needs remains in question. Nonetheless, algal-based TGBs are an intense focus within the alternative fuels research community and production companies are paying close attention to advances. Of particular interest are microalgae-based TGB products. Technical and economic aspects of high-yield production of biodiesel from microalgae are currently being studied by many groups. A few review papers have discussed the processes and challenges in detail [17,18,19,20]. The potential of ethanol production from algae has also been investigated [21,22,23,24]. In both cases, it is generally agreed that there is significant potential for algal-based products; however, progress toward commercial-scale biofuels production from algae has been slow due to multiple challenges in production and processing.

Whether in open ponds or in closed photobioreactors, culturing algae requires consideration of numerous environmental conditions. Environmental factors such as temperature, light, pH, and nutrients not only affect photosynthesis and growth rate of the algae, but also influence the activity of cellular metabolism and composition. These effects have been identified individually by researchers [25,26,27,28,29,30,31], however, no consolidated review on the effect of these factors on algae is available. Most of the recent reviews have focused on production and processing techniques [11,16,32] with few data on the impacts of environmental and nutritional factors on algal growth rates. One recent, highly informative review, does consider microalgae and multiple environmental factors including temperature, light, pH, and salinity [33,34,35], however, it is almost exclusively focused on impacts of environmental factors on lipid production. Understanding how these factors influence algal growth and broader metabolic functions (not only lipid induction) is critical for successful scale-up of algae cultures in commercial systems for algal biofuels and bioproducts production. This review directly addresses this issue by focusing on the impacts of environmental and nutritional factors on algae biochemistry specifically related to biofuels production processes. The paper is organized as follows: This introductory section is followed by a brief overview of different production and processing technologies. The subsequent section provides a detailed review of the environmental and nutritional factors that affect algal growth. The third section addresses interaction effects between environmental factors and nutrient availability. This is followed by a discussion of sustainable production of algal biofuels. Finally, we offer a discussion of future outlook and some concluding remarks.

1.1. Algae Production, Processing and Use

Algal production processes can be categorized into three general classes of growth regimes based on the energy source and mode of utilization; these include: photoautotrophic, heterotrophic and mixotrophic. These growth processes can occur in open raceway ponds or closed bioreactor systems [36]. Different production schemes involving combinations of different growth regimes in various reactor configurations have been proposed in an effort to maximize biofuel productivity. Substrates for algal feedstock production can range from industrial effluents and municipal waste water to synthetic media consisting of sugars from molasses, starch or lignocellulosic feedstocks depending on the growth regime and bioreactor configuration. Although open raceway ponds require low energy inputs and lower capital costs, several issues such as contamination (i.e., by unwanted algal species as well as viral, bacterial and fungal pathogens) and low final biomass concentrations (often less than 1.0 g/L) increase production costs to levels that are still economically unviable for large-scale production of biofuels [32]. Closed photobioreactors (CPBR) in different configurations have been proposed to address the contamination issues associated with open raceway systems. CPBRs also permit more stringent control of growth conditions and harvesting [13]. However, CPBRs often require extensive upfront capital investments compared to open raceway ponds and therefore face similar commercialization challenges.

In addition to system setup, there are challenges associated with specific stages within the production process. Lipid and starch fractions of algae can be processed into renewable fuel by several different processing methods, including: pyrolysis [37], thermochemical liquefaction [38], fermentation [39], and transesterification [40]. Most thermochemical conversion methods such as direct combustion, gasification, and pyrolysis require low-moisture-content biomass. This poses a challenge due to high energy requirements for drying algal feedstock. However, hydrothermal liquefaction process can use wet slurry for algal-oil production [41] thus reducing costs associated with drying. Biochemical conversion processes such as anaerobic digestion produce methane (60%–70%) and carbon dioxide (30%–40%) [42]. Fermentation of sugars produced from the starch fraction can be used to produce ethanol from algal biomass [39]. Advantages and limitations of these processes are tabulated in Table 1.

Table 1. Advantages and limitations of biofuel production processes from microalgae.

Process	Advantages	Limitations
Pyrolysis	High bio-oil yields possible(up to 57.5% w/w for fast and flash pyrolysis [43])	Low-moisture-content biomass required High-energy-content required to dry feedstock
Thermochemical liquefaction	Algal wet slurry can be used Energy (and cost) reduction High yields possible (up to 60% w/w [44])	Reactors are complex and expensive
Fermentation	Co-products can be utilized Conversion of sugar to bioethanol possible	Long processing times required Biomass has to be preprocessed to be converted to sugars
Transesterification	Enhanced physical properties of renewable fuels Biodiesel has a current market that simplifies commercialization	Limited to conversion of lipids and does not utilize carbohydrate and protein fractions of feedstock

Details concerning algae biomass production processes, different processing technologies for conversion into biofuels and bioproducts, and challenges associated with commercialization of algal biofuels have been discussed thoroughly in recent reviews [11,32,35,45]. A summary of various algae production, harvesting and processing alternatives are presented in Figure 1 [36,46,47].

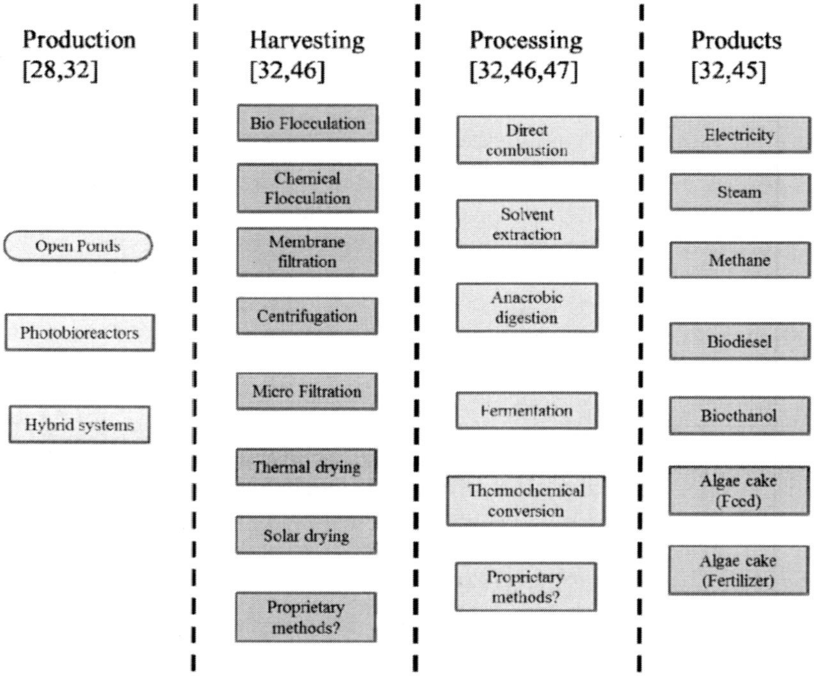

Figure 1. Production, harvesting and processing alternatives for algae.

1.2. Relevance of Individual Algal Cellular Components for Biofuels Production

The exploitation of microalgae as a protein source has led to increased interest in the use of microalgae (e.g., Spirulina, Chlorella and Scenedesmus) in health food production [48]. Algae such as Dunaliella and Spirulina are also used for pigment production [49,50]. However, it is the high density lipid content that renders algae (e.g., Chlamydomonas, Nannochloropsis, Schizochytrium, Chlorella, Nitzschia) attractive for biofuels production [51]. Lipid content in algae can be as high as 50% (dry-weight) with higher percentages of total dry weight (~60%) found as protein and starch [11]. These fractions of microalgae can be used to produce biodiesel [20,52], bioethanol [22,23], biohydrogen [53,54], bioplastics [55], and other products, while simultaneously contributing to CO_2 mitigation [56,57]. Still, it is the lipid fraction that has been the major focus for biodiesel production and renewable jet fuel. Depending on the species and growth conditions, dry algae weight can contain 20%–50% lipid [52], which is the major substrate for biofuels production. Total lipid composition includes (but is not limited to) neutral lipids, polar lipids, wax esters, sterols, hydrocarbons, and prenyl derivatives such as tocopherols, carotenoids, terpenes, quinones and phytylated pyrrole derivatives, including chlorophylls [51].

Carbohydrates (primarily starch) are another valuable component of the algal cell. Typical dry weight content of carbohydrates in algae range from 20% to 40% of total cell mass [58]. Certain components, such as starch, can be easily converted to ethanol by hydrolysis and fermentation. Currently, the starch used for ethanol production is obtained from food crops such as corn, wheat, sorghum, and others [1]. For algae, high starch strains including Chlorella vulgaris (with 37% dry weight starch) [59], are being studied for potential use has high-yield feedstocks. Algal starch is known to be readily fermentable by yeast [60] and, therefore, is being intensely studied for use in ethanol production [22,23].

Upon utilization of lipids and starch fractions for bio-oil production, the residual algae cake, which is rich in proteins, is important for producing valuable co-products. Protein (and starch) content can constitute up to 60% of dry weight of algae [48]. This residual protein from the biomass can be used for livestock, poultry, and fish feed additives [11]. It has been reported that algae can replace about 5%–10% of conventional protein sources in poultry feed [61]. Recently residual algae cake after lipid extraction has been used in large animal feeding trials [62]. However high concentrations of nucleic acids in algae can pose challenges for the utilization in animal feed applications [63].

2. EFFECT OF ENVIRONMENTAL FACTORS

During photosynthesis, using only light and nutrients, algae produce lipids, proteins, and carbohydrates. The relative amounts of these metabolic products are tightly linked to environmental and nutrient conditions including: the amount and intensity of sunlight; CO_2 levels; pH; temperature; available nutrients; and, the presence (or absence) of other organisms. Carbon, hydrogen, and oxygen are required non-mineral nutrients for algal growth. Macronutrients include nitrogen, phosphorus, sulfur, potassium and magnesium. Micronutrients such as iron and manganese are also required in small amounts (30–2.5 ppm) while other elements such as cobalt, zinc, boron, copper and molybdenum are essential trace elements (4.5–2.5 ppm) [64]. Collectively, environmental conditions (especially light and temperature) and the availability of non-mineral nutrients, macronutrients, and micronutrients, greatly influence the biochemical composition of microalgae [25,30,65,66,67]. Other factors such as pH and the present toxic metals are also important factors impacting algal growth and metabolism. In general, all of these factors can affect photosynthesis, thus altering carbon fixation and the allocation of carbon into different types of macromolecules. In turn, the cell's macromolecular composition determines its usefulness in biofuels production.

2.1. Temperature

Temperature is perhaps one of the most important environmental factors that influences algal growth rate, cell size, biochemical composition and nutrient requirements. In the United States, algae grow under a broad range of temperatures (from 15 to 40 °C), depending upon strain, region, and season. Below optimal growth temperatures, growth rate (μ) increases with increasing temperature but declines markedly above the species- or strain-specific optimum [68]. Growth at temperature optima results in minimal cell size [69,70] and the efficiency of carbon and nitrogen utilization decreases at non-optimal temperatures [71]. It has been suggested that changes in cytoplasmic viscosity under sub-optimal temperature conditions is responsible for less efficient carbon and nitrogen utilization [72,73]. Temperature may also play a key role in photoinhibition, which is known to impact algal growth rate. Several mechanisms of temperature-dependent photoinhibition have been postulated. These include mechanisms under which: (i) low temperature results in reduced electron transport at a given photon flux rate due to slower rate of CO_2 fixation; (ii) low temperature inhibits the active oxygen species, which results in reducing photoinhibition by protecting PSII; and (iii) low temperature inhibits the synthesis of the D1 protein degraded during photoinhibition, consequently impeding the PSII repair cycle [74].

One of the most commonly observed changes with temperature shift is the alteration in the level of unsaturation of fatty acids in the lipid membrane [75,76,77]. In a study on eight marine plankton species, fatty acids (14:0) increased from ~4% at 10 °C to >20% at 25 °C while PUFA (polyunsaturated fatty acids) were consistently higher at lower temperature (10 °C) [27]. Dunaliella salina has shown a considerable increase in fatty acid unsaturation in response to decrease in temperature from 30 to 12 °C [78]. Lower temperatures decrease the fluidity in the cell membrane. Cells then compensate by increasing levels of unsaturated fatty acids to increase fluidity. However, it also makes the membranes more susceptible to damage by free radicals [73,79]. Along with greater fluidity, increased levels of unsaturated fatty acids tend to enhance the stability of the cellular membranes (particularly the thylakoid membrane). This, in turn, protects the photosynthetic machinery from photoinhibition at low temperatures [79]. For example, in a study involving Botryococcus braunii, a green alga that secretes extracellular lipids, differences in lipid composition were observed at three different growth temperatures (18 °C, 25 °C, and 32 °C). Intracellular lipid synthesis was found to be inhibited at supra-optimal temperature (32 °C); consequently, lipid content decreased to 5% dry weight at 32 °C in comparison with 22% at 25 °C. The decrease in lipid content led to an accumulation of polysaccharides. However, temperature did not affect the secretion of extracellular lipids [66]. Similar effects were observed in Nannochloropsis oculata and Chlorella vulgaris, both of which have an optimum growth temperature of 25 °C. Increasing the growth temperature from 20 to 25 °C doubled the lipid content (from 7.90% to 14.92%) in N. oculata. Increase of temperature from 25 to 30 °C decreased the lipid content in C. vulgaris from 14.71% to 5.90% [80].

Increasing temperature beyond the optimum reduces protein synthesis and consequently results in decreased growth rates [81]. Morris et al. [25] studied the growth of alga, Phaeodactylum tricornutum, a marine diatom, and reported a considerable increase in protein synthesis rates at night with lower the temperatures, presumably due to the fact that protein synthesis is a significant component of nighttime algal metabolism [82]. Similarly, Rhee and Gotham [83] observed an increase in protein concentration in Scenedesmus sp. with decreasing temperature. However, in this study, the efficiency of protein synthesis (in terms of rate of protein synthesis per unit RNA) was reduced upon increasing temperatures beyond the optimum. An increase in temperature from 20 to 30 °C in cultures of Ulva pertusa resulted in higher intercellular free amino acid concentrations from approximately 840 to 1810 mg/100 g dry weight [84]. An increase in free amino acid concentration is an indicator of lower protein content.

Temperature is also reported to impact starch content in the algal cell. Starches are synthesized by phosphorylated metabolites in the dark reactions of the photosynthesis cycle using energy-rich phosphate bonds (i.e., ATP) formed in the light reactions [85]. Increased temperature leads to degradation of the starch produced [65,86]. Enzymes that have been suggested to play a critical role in the temperature dependent degradation of starch are α-amylase and α-glucan phosphorylase [87]. Nakamura and

Miyachi studied the effect of temperature on the starch degradation in Chlorella vulgaris grown autotrophically at 20 °C [65]. This study reported a significant reduction in starch (17%) with concomitant increase in sucrose (57%), when culture was exposed to 38 °C for 10 min after 30 min at 20 °C. However, another study with Chlorella vulgaris grown autotrophically at 38 °C, reported a reduced degradation of starch at 38 °C [88]. This effect could be explained by noting that the starch produced at 38 °C was not subjected to any degradation since there was no change in temperature, but the cultures grown at 20 °C degraded the starch into sugars in response to increasing culture temperatures. Consistent with this reasoning, Nakamura and Imamura observed a reversible transformation of high-molecular-weight L starch (amylopectin-like molecule fraction with MW > 2×10^6) to low-molecular-weight S starch (amylose-like molecule fraction with MW <10^4) at high temperatures (38 °C) [89].

Temperature has a significant effect on the formation of carotenoids. Carotenoids absorb light energy for use in photosynthesis. They also protect chlorophyll from photodamage [90]. Furthermore, they play a vital role in the photosynthetic reaction center by either participating in the energy-transfer process or protecting the reaction center from auto-oxidation. Carotenoid accumulation in algal species increases with temperature because of the increased oxidative and photodamaging effects noted at elevated temperatures [91,92,93]. Tjahjono et al. [91] reported a three-fold increase in astaxanthin formation in the green alga Haematococcus pluvialis with an increase in cultivation temperature from 20 to 30 °C. These results were confirmed by another study on a different green alga, Chlorococcum sp., in which a two fold increase in total carotenoid content was observed by raising the temperature from 20 to 35 °C under conditions of nitrogen deprivation [92]. Increase in carotenoid formation with increasing temperature is generally attributed to cellular response to enhanced active free oxygen radical formation [91] or increased biosynthetic enzyme activity [92].

2.2. Light

Light is the energy source during photoautotrophic growth phase and organisms use light energy to convert carbon dioxide to organic compounds—especially, sugars. The range of light intensity in USA varies from 1500 to 8500 W℘h/m^2/day with strong regional and seasonal dependence [94]. Light intensity effects growth of algae through its impact on photosynthesis [95]. Although rate of growth under increasing light intensity is a function of strain and culture temperature, the growth rate of algae is maximal at saturation intensity and decreases with both increase or decrease in light intensity [96]. The photoadaptation/photoacclimation process in algae leads to changes in cell properties according to the availability of light and an increase in photosynthetic efficiency [97]. Adaptation can occur through multiple mechanisms such as changes in types and quantities of pigments, growth rate, dark respiration rate or the availability of essential fatty acids [67]. Morphological photoacclimation is accompanied by changes in cell volume and the number and density of thylakoid membranes [98]. Algae overcome light limitation by desaturation of chloroplast membranes [99]. Light intensity increase above saturating limits causes photoinhibition [30,100]. This is due to the disruption of the chloroplast lamellae caused by high light intensity [101] and inactivation of enzymes involved in carbon dioxide fixation [102]. For example, growth rate of Dunaliella viridis decreased to 63% with increase in light intensity from 700 to 1500 $\mu mol \cdot m^{-2} \cdot s^{-1}$ [30].

Light intensity also affects the cellular composition of algae. Dunaliela tertiolecta exhibits a decrease in protein content and an increase in the lipid fraction with increasing light intensities up to saturation [82]. Similar results were reported by Morris et al. [25], in a study on the marine diatom, Phaeodactylum tricornutum, in which low light (400 lux at the culture surface) led to an increase in the rate of protein synthesis. Low light intensity has been observed to result in higher protein content while high photon flux density (PFD) results in increased extracellular polysaccharide content [102]. Absence of light was observed to increase the total lipid content of the D. virdis but reduce triglycerides, free fatty acids, free alcohols and sterols [103]. In Nannochloropsis sp., grown under low light conditions (35 $\mu E \cdot m^{-2} \cdot s^{-1}$), 40% of the total lipids were found to be galactolipids and 26% were found to be triacylglycerols. In the same system, high light (550 $\mu E \cdot m^{-2} \cdot s^{-1}$) conditions resulted in an increased synthesis of triacylglycerol with a reduction in galactolipid synthesis [26]. High light, in general, leads to oxidative damage of PUFA. Numerous studies have suggested that the cellular lipid content and PUFA decrease with increase in light intensity [104,105,106]. Conversely, Nannochloropsis cells under low light conditions were characterized by high lipid content and high proportions of eicosapentaenoic acid (EPA; 5,8,11,14,17-icosapentanoic acid) [26]. Confirming this observed trend, another study on the same species reported an increase in unsaturated fatty acids mainly due to an increase in EPA (from 44.3% to 60.7% of the organic content) and a decrease in protein content, with decreasing irradiance [67]. Increase in PUFA under light-limited growth conditions are coupled with an increase in total thylakoid membrane in the cell [98]. However, there are some contradictory studies in which PUFA levels were observed to be increasing with higher light intensity [107]. This difference in response to environmental conditions by different alga may be related to difference in their metabolic pathways. Increase in oxygen-mediated lipid desaturation could be one potential reason for the observed increase in PUFA levels under conditions of higher light intensity [107].

In addition to total light intensity, light cycles and the spectral composition of incident light impact algae. For example, Wu and Merchuk [108] investigated the effect of light and dark cycles on the growth of algae and observed that with increasing photon flux density (PFD), specific growth rate increases up to a certain threshold PFD value after which a decline in growth rate was observed. Sustained high light intensities have also been reported to cause photoinhibition and reduce light utilization efficiency. Light utilization efficiency may be optimized by prolonging the dark period under conditions of high light intensity. This allows the photosynthesis machinery in the cell to fully utilize captured photons and convert them into chemical energy thus avoiding the effects of photoinhibiton [109].

Since the energy content of near-ultraviolet (300–400 nm) and blue light (400–480 nm) is greater than that of red light (620–750 nm), fewer photons of blue light are required to achieve an equivalent magnitude of energy intensity using red light. In addition to differences in the energy intensity, specific components of light are known to impact the cellular regulatory processes including: chlorophyll synthesis, photodamage repair, and cell division. For example, blue light was shown to be essential for the division of Chlamydomonas reinhardtii cells [110]. Evidence for the influence of blue light on short-term growth rate is equivocal [111]. It has been observed that blue and red light can help to increase growth and polysaccharide production [100]. Emerson and Lewis also reported blue and red light to be the most effective for photosynthesis of Chlorella [112]. Miyachi and Kamiya studied the starch formation in Chlorella vulgaris under blue (456 nm) and red (660 nm) light. They reported that the carbon pathway in photosynthesis is regulated by short wavelength light (blue), even under low intensity [113]. Red light of high intensity was observed to incorporate carbon from CO_2 into sucrose and starch synthesis pathways. However, superimposition of monochromatic blue light even at low intensities resulted in a significant decrease in sucrose and starch formation along with increasing levels of alanine, aspartate, glutamate, glutamine, malate, citrate, lipids and the alcohol-water-insoluble non-carbohydrate fraction [113].

Ultraviolet light (UV; 215–400 nm) adversely affects the algal primarily due to the damage to the photosynthetic machinery in the cells. UV-B (215–380 nm) causes more damage to the cells compared to UV-A radiation (380–400 nm) even at similar intensities [114]. The UV-B radiation causes direct damage to cellular DNA, UV-A damage is limited to indirect damage through enhance production of reactive oxygen and hydroxyl radicals. At moderate levels, UV-A may stimulate photosynthesis while UV-B has a negative effect of photosynthesis irrespective of the intensity. Some of the response of the algae to minimize the damage caused by UV radiation includes migration, development of protective cell walls, increased synthesis of carotenoids and other pigments [114,115,116].

2.3. pH

One of the most important factors in algal cultivation is pH since it determines the solubility and availability of CO_2 and essential nutrients, and because it can have a significant impact on algal metabolism [28,117]. Due to uptake of inorganic carbon by algae, pH can rise significantly in algal cultures [118]. Maximum algal growth occurs around neutral pH, although optimum pH is the initial culture pH at which an alga is adapted to grow. Changing pH in media may limit algal growth via metabolic inhibition [119]. Pruder and Bolton observed that T. pseudonana cells adapted to low pH (6.5) had lower growth rate at sub-optimal pH (8.8) [120]. Normal growth rate was restored after the pH was lowered by addition of HCl. Similar results were reported by Chen and Durbin, where photosynthetic rate and algal growth was minimal at pH 9.0, but carbon uptake rates were enhanced when the pH was lowered to 8.3 [28].

Notably, pH is the major determining factor of relative concentrations of the carbonaceous species in water [121]. Higher pH limits the availability of carbon from CO_2, which, in turn, suppresses algal growth [28,121]. At higher pH, the carbon for algae is available in form of carbonates [122]. Higher pH also lowers the affinity of algae to free CO_2 [121,123]. In photoautotrophic cultures, replacement of CO_2 taken up for photosynthesis is slower resulting in a decrease of CO_2 partial pressure and thus leading to an increase in pH [120]. Alkaline pH increases the flexibility of the cell wall of mother cells, which prevents its rupture and inhibits autospore release, thus increasing the time for cell cycle completion [124]. Alkaline pH indirectly results in an increase in triglyceride accumulation but a decrease in membrane-associated polar lipids because of cell cycle inhibition. Membrane lipids in Chlorella were observed to be less unsaturated under conditions of alkaline pH [124].

Similar to alkaline pH, acidic conditions can alter nutrient uptake [125] or induce metal toxicity [126,127] and thus affect algal growth. As previously stated, most species of algae grow maximally around neutral pH (7.0–7.6). This has been observed in studies of Ceratium lineatum, Heterocapsa triquetra and Prorocentrum minimum [118] and Chlamydomonas applanata [31]. Visviki and Santikul [31] studied the growth of

Chlamydomonas applanata within a pH range 1.4 to 8.4 with 1 point increments. No growth was observed from pH 1.4 to 3.4, above which tolerance of pH in C. applanata was observed (with optimum growth observed at 7.4). Exponential growth was observed for up to five days at pH 5.4 to 8.4, but maximum growth was achieved at pH 7.4. In a study on Chlamydomonas acidophila at pH 4.4, it was observed that hydrogen ions denature V-lysin, a proteolytic enzyme that facilitates releasing of daughter cells from within the parental wall [128]. Hargreaves and Whitton [129] studied the effects of low pH on the morphology of five algal species. Acidic conditions (pH 1.3–1.5) were observed to limit the motility of cells in Chlamydomonas applanata var. acidophila and Euglena mutabilis. Coleman and Colman studied the effect of external pH on photosynthesis of Coccochloris peniocystis and found a significant decrease in total accumulated carbon and oxygen evolution at pH 5.0 and 6.0, which suggested the reduction in photosynthesis in this cyanobacteria (blue-green alga) at these pH ranges [130]. Maintenance of neutral intracellular pH in an acidic pH external environment would require an expenditure of energy to pump protons out of the cell [131]. On the other hand, acid-tolerant algae such as Chlorella saccharophila [132] and Euglena mutabilis [133] can change intracellular pH in response to changing external pH. In Chlorella saccharophila, an internal pH of 7.3 was maintained for an external pH range of 5.0–7.5; however, decreasing the pH further to 3.0 caused a decrease in cellular pH to 6.4 [132]. Similarly, Euglena mutabilis exhibited an internal pH range from 5.0 (at low external pH < 3.0) to 8.0 (at high external pH > 9.0) [133]. The energy required to maintain internal pH in these acid-tolerant algae is conserved as the internal pH goes down. This may be a mechanism for maintaining cellular metabolism such that algal growth is not drastically affected under acidic conditions [132]. Such a mechanism would endow acid-tolerant algae with the ability to adjust internal pH in response to external pH fluctuations, thereby, maintaining an energy advantage over acid-intolerant species at low external pH.

Some algae such as Dunaleilla acidophila adapt to acidic conditions in growth media by accumulating glycerol to prevent the osmotic imbalance caused by high concentrations of H_2SO_4 [134] while other species such as Chlamydomonas sp. and Pinnilaria braunii var. amplicephala (an acidophilic diatom) accumulate storage lipids such as triacylglycerides under highly acidic conditions (pH 1) [135]. Another adaptation observed under acidic conditions is an increase in saturated fatty acid content, which reduces membrane fluidity and inhibits high proton concentrations [135]. Such adaptation was reported in a Chlamydomonas sp., in which total fatty acid content increased from 2% at pH 7 to 2.4% at pH 2.7, a modest but statistically significant increase [136].

Under alkaline conditions whereby the extracellular pH is higher than intracellular pH, the cell must rely on active transport of HCO_3^- and not on passive flux of CO_2^- for inorganic carbon accumulation [121,137]. Affinity of algae for CO_2 increases at lower pH [121,123]. Moroney and Tolbert studied the effects of pH on carbon uptake of Chlamydomonas reinhardtii [138]. They reported an efficient utilization of CO_2 for photosynthesis at lower pH (<6.95). However, at high external pH (6.95–9.5), where HCO_3^- dominates, algae cannot efficiently accumulate carbon and require high supply of carbonates for maintaining photosynthetic activity. In more acidic environments, where the internal pH exceeds that of the surrounding medium and where CO_2 comprises a major portion of the total external inorganic carbon, carbon accumulation is thought to be accomplished by the passive movement of CO_2 along a pH gradient into the cell or chloroplasts [139,140].

2.4. Salinity

Salinity is another important factor that alters the biochemical composition of algal cells (salinity refers primarily to sodium chloride concentration unless otherwise specified). Exposing algae to lower or higher salinity levels than their natural (or adapted) levels can change growth rate and alter composition. For example, higher salinity increases the algae lipid content [141,142,143,144]. Dunaliella, a marine alga, exhibited an increase in saturated and monounsaturated fatty acids with an increase in NaCl concentration from 0.4 to 4 M [145]. In another study with Dunaliella tertiolecta, an increase in intracellular lipids (60% to 67%) and triglyceride concentration (40% to 56%) with an increase in NaCl concentration from 0.5 (freshwater concentration) to 1.0 M was observed [146]. Increasing the NaCl level in cultures of Botryococcus braunii, a fresh water alga, showed an increase in growth rate, carbohydrate content, and lipid content; however, the greatest biomass concentration was achieved at the lowest salinity level [147]. These results are supported by another study in which lipid content of Botryococcus braunii grown in 0.50 M NaCl was higher compared to media without NaCl addition, but protein, carbohydrates, and pigments levels were lower [143]. Another study with the same alga reported a decrease in protein content with unchanged carbohydrate and lipid content with an increase in salinity [148]. This study also reported reduced growth at higher salinities, which may have been due to an inability of the alga to adapt to high salinity. A study with Tetraselmis suecica also reported reduction in protein content per cell of up to 20% with increasing salinity [144].

2.5. Nutrients

Considerable variation in the biochemical composition under conditions of nutrient limitation can be observed in algae depending upon which nutrient is limited and to what degree. In general, the growth rate of algae is proportional to the uptake rate of the most limiting nutrient under optimal conditions of temperature and pH and is generally described by Michaelis-Menten equation [149].

Nitrogen and phosphate are two important macronutrients for growth and metabolism of algal cells. Nitrogen is a fundamental element for the formation of proteins and nucleic acids. Being an integral part of essential molecules such as ATP, the energy carrier in cells, phosphate is another very important nutrient. Phosphate is also a part of the backbone of DNA and RNA, which are essential macromolecules for all living cells. Phosphorus is also a key component of phospholipids. It is not unusual for algae to become nutrient-limited (i.e., nitrogen- and phosphorus-limited) in the natural environment [70]. Limitation of these key nutrients shifts the metabolic pathway of the organism. For example, nitrogen and phosphorus starvation shifts the lipid metabolism from membrane lipid synthesis to neutral lipid storage. This, in turn, increases the total lipid content of green algae [58]. Specific effects of major nutrients are discussed below.

Carbon. Carbon, hydrogen, and oxygen are three essential non-mineral nutrients. Abundance of hydrogen and oxygen in the media for algae cultures means that their availability is not a challenge to cellular growth or metabolism. Carbon is one of the other major nutrients that must be supplied. It is essential for photosynthesis and hence algal growth and reproduction. Carbon fixed by the algae can end up in three destinations; it will either be used: (a) for respiration; (b) as an energy source; or, (c) as a raw material in the formation of additional cells [150]. Reduced carbon fixation rate implies a reduction in algal growth rate. Algae require an inorganic carbon source to perform photosynthesis. Carbon can be utilized in the form of CO_2, carbonate, or bicarbonate for autotrophic growth and in form of acetate or glucose for heterotrophic growth. CO_2 in water may be present in any of these forms depending upon pH, temperature and nutrient content:

$$CO_2 + H_2 \leftrightarrow H_2CO_3 \leftrightarrow H^+ + HCO_3^- \leftrightarrow 2H^+ + CO_3^{2-} \quad (1)$$

With an increase in pH, carbonate increases while molecular CO_2 and bicarbonate decrease [28]. At the average pH of seawater (8.2), 90% of the total CO_2 is present in the form of HCO_3^-; only 1% exists as molecular CO_2 and the rest is bicarbonate [122,151].

Riebesell et al. [152] studied the effect of CO_2 concentration on lipid distribution in Emiliania huxleyi. A significant effect of CO_2 on the composition of the polyunsaturated fatty acids and alkenones was reported. Specifically, lower CO_2 concentrations led to an increase in 22:6 (n−3) PUFA, whereas 14:0 fatty acids were found to be predominant at higher CO_2 concentrations. Along with the change in composition, increased CO_2 was also observed to increase the amount of fatty acid accumulation in Dunaliella salina [153]. Similar increase in fatty acid content and unsaturation with increase in CO_2 concentrations were reported by Tsuzuki et al. [154]. Another study with the cyanobacterium Spirulina platensis reported that elevated CO_2 concentrations decrease relative concentrations of proteins and pigments in the cells but increase carbohydrate content. This change in the cell composition was accompanied by reduction in the maximum biomass yield [155].

Nitrogen. Nitrogen is an essential constituent of all structural and functional proteins in the algal cells and accounts for 7%–20% of cell dry weight [58]. Inorganic nitrogen taken up by algae is rapidly assimilated into biochemically active compounds and recycled within cells to meet changing physiological needs [156,157]. Major effects of nitrogen deficiency in algal culture include the enhanced biosynthesis and accumulation of lipids [75,80,158,159,160] and triglycerides [161,162] with a concomitant reduction in protein content [25,29,163,164,165,166]. This, in turn, results in a higher lipid/protein ratio [80] at the expense of growth rate [167]. Thus, attempts to increase lipid concentration via nitrogen limitation must be carefully evaluated to ensure high lipid productivity [168]. Algae grown in nitrogen-depleted cultures also tend to divert their photosynthetically fixed carbon to carbohydrate synthesis [58]; however, the physical significance of this is not clear. Other effects of nitrogen reduction include decrease in oxygen evolution, carbon dioxide fixation, chlorophyll content, and tissue production [169,170]. Holm-Hansen et al. [164] reported an increase in amino acid content of Chlorella pyrenoidosa at the expense of sugar phosphates (such as glucose-6-phosphate, fructose-6-phosphate) with addition of ammonium (nitrogen source) to the growing culture.

Degradation of phycolbillisomes with nitrogen limitation has been demonstrated in the case of cyanobacteria and red alga [171]. Phycolbillisomes are the light harvesting antennae of photosystem II in these algae. Photosynthesis continues at a reduced rate, until cell nitrogen falls below a particular species-dependent threshold value. Under nitrogen deficient conditions, Spiriluna platensis cells exhibit reduced carbon fixation capacity even under normal to high available CO_2 concentrations [155]. Nitrogen starvation also alters the enzyme balance of cells, resulting in the synthesis of lipids and a decrease in chlorophyll synthesis leading to excess carotenoids in the cells [172]. Dunaliella sp. and Haematococcus pluvialis are observed to accumulate high amounts of carotenoids, astaxanthin and its acylesters (up to 13% w/w), when grown under nitrogen-depleting conditions [173,174,175]. Zhekisheva et al. [176] reported that under nitrogen depleting conditions, Haematococcus pluvialis produced fatty acids and astaxanthin in a 5:1 ratio. It was suggested that the production of the oleic acid-rich triacyl-glycerols and the esterification of the astaxanthin, maintain a high content of astaxanthin esters by enabling the oil globules.

Phosphorus. Phosphorus is an important component required for normal growth and development of algal cells [58]. It has been shown that phosphorus, rather than nitrogen, is the primary limiting nutrient for microalgae in many natural environments [177]. Phosphorus typically constitutes 1% of dry weight of algae [178], but it may be required in significant excess since not all added phosphate is bioavailable due to formation of complexes with metal ions [52]. Immediate effects of phosphorus limitation include a reduction in the synthesis and regeneration of substrates in the Calvin-Benson cycle and a consequential reduction in the rate of light utilization required for carbon fixation [179].

Phosphorus limitation also leads to accumulation of lipids. Total lipid content in Scenedesmus sp. was observed to increase from 23% to 53% with a reduction in initial total phosphorus (as phosphate) concentration of 0.1 from 2.0 mg L^{-1} [180]. Phosphatidylglycerol (PG), which is one of four major glycerolipids constituting membrane lipids in chloroplasts, was observed to decrease with phosphorus limitation in Chlamydomonas reinhartdtii [181]. PG is essential for cell growth, the maintenance of chlorophyll-protein complex levels, and normal structure-function of the PSII complex. Total acidic lipid (such as sulphoquinovosyldiacylglycerol and PG) content of the chloroplast did not change significantly since a decrease in one acidic lipid was accompanied by an increase in another acidic lipid [181]. Phosphate limitation also reduces the synthesis of n-3 PUFA [182].

Similar to the effects of nitrogen deficiency, phosphorus starvation reduces chlorophyll a and protein content thereby increasing the relative carbohydrate content in algal cells [29,183,184]. Phosphate deficiency has been demonstrated to result in accumulation of astaxanthin and an overall reduction in cell growth [185]. A decrease in cellular phycobilisome under conditions of phosphorus deficiency (due to cell division and the cessation of phycobilisomes synthesis) has also be shown [171]. Theodorou et al. [186] observed that phosphorus starvation in Selenastrum minutum reduces respiration rate.

Trace Metals. Trace metals are metals present in algal cells in extremely small quantities (<4 ppm) but that are an essential component of phycophysiology. Iron (Fe), manganese (Mn), cobalt (Co), zinc (Zn), copper (Cu) and nickel (Ni) are the six most important trace metals required by algae for various metabolic functions [187]. As the aqueous concentration of trace metals is not an indicator of the bioavailability of metals, trace metal availability to algae is highly dependent on speciation (free ion concentration) [188]. Deficiencies in trace metals can limit algal growth, whereas excesses or high metal concentrations (above the toxicity threshold) may inhibit growth, impair photosynthesis, deplete antioxidants, and damage the cell membrane.

Iron is an important trace metal for normal growth and functioning of photosynthesis and respiration in algae. It acts as redox catalyst in photosynthesis and nitrogen assimilation and mediates electron transport reactions in photosynthetic organisms [131]. Iron limitation significantly depresses photosynthetic electron transfer, resulting in a reduction in NADPH formation. Reduction in iron decreases the cellular abundance of ferredoxin, which contains Fe, and forces the substitution of flavodoxin, a non-iron functional equivalent, in the cell [189,190,191,192]. Since the catalytic capacity of ferredoxin is much higher than flavodoxin, this can be problematic [193]. Iron limitation also reduces cellular chlorophyll concentration [194]. High concentrations of iron in cultures of Chlorella vulgaris were observed to increase the lipid content [195]. Decrease in iron content reduces carotenoid composition [185,196].

While there are some non-essential metals (e.g., Cd, Pb and Cr), which can inhibit many metabolic processes even in small quantities [197], there are some essential elements (e.g., Zn and Cu) which when in excess can cause toxicity [198]. The cell surfaces of algae contain a number of functional groups with high affinity for metal ions and that carry a net negative charge mainly due to carboxylic, sulfhydryl, and phosphatic groups [199]. These groups are binding sites that transport metal ions across the cell membrane and into the cell. Cu, Ni and Fe are metals that are commonly observed to be toxic to the algae, if present in supra-threshold concentrations. Cu is one of the most toxic of these metals. Toxic metals can inhibit carbon fixation and delay nutrient uptake [200]. Copper ions were observed to inhibit both cell division and photosynthesis in Asgerionella glacialis (marine diatom) and Chlorella pyrenoidosa (freshwater alga) [201]. Metal toxicity in algae is observed to be a pH-dependent effect, possibly due to the "competition between H^+ and free metal cations for cellular binding sites". For example, copper toxicity was observed to increase 76-fold from pH 5.0–6.5 for the green alga Scenedesmus quadricauda [202]. Cadmium (Cd) is particularly toxic to algal cells. Although Cd has no biological significance in a living cell, it is taken up by marine cells in the form of complexes with organic matter and is absorbed onto organic matter and inorganic matter in ionic form [203]. Cd is inhibits phosphorus uptake, which is also a pH-dependent phenomena. This toxic effect increases in the pH range of 5.5–8.5 [202]. Zinc is also a toxic metal which is rapidly taken up by the algae and is incorporated primarily into polysaccharide and nucleic acid fractions [204].

3. INTERACTION AMONG ENVIRONMENTAL FACTORS

Environmental factors may influence other systems factors related to algal growth or cellular composition. For example, increase in temperature can lead to reductions in nutrient availability [205]. Such interaction effects have been observed in many studies and could underlie contradictory results between some studies. For example, in a study on phytoplankton in Antarctic Ocean, Smith and Morris [206] reported that at higher temperatures phytoplankton incorporate more carbon into the protein fraction with a concomitant reduction in lipids. To the contrary, Morris et al. [25] reported an increase in the protein fraction of P. tricornutum at low temperatures. This variability could be due to the species-specific effects, differences in light intensity, and/or differences in the primary growth conditions.

In another study, Kudo et al. [207] observed the effect of iron stress on growth rate and cellular composition of the marine diatom Phaeodactylum tricornutum, over the temperature range of 5–30 °C. (Note that the optimum temperature for growth of P. tricornutum was 20 °C). The growth rate of Fe-stressed cells was 50% of Fe-replete cell growth rate at the optimum growth temperature. Differences in growth rates diverged significantly at suboptimum temperatures. It was also reported that at optimal temperature, the C:N ratio in the cell decreased by about 5% in cells induced with an iron stress (2 μM to 2 nM Fe). However, an increase of about 4% was demonstrated with the same transition in iron concentration but at a lower temperature (10 °C).

Algal growth rates are also affected by light-by-nutrient interactions. Cloern et al. [208] developed a model of phytoplankton chlorophyll—specifically, carbon ratio as a function of light and nutrients. The model suggests an increase in growth efficiency with nutrient availability under low light conditions. Morgan and Kalff [209] studied the interaction of temperature and light on the growth of Cryptomonas erosa under nutrient saturated conditions. Results indicated that algal carbon uptake capacity is reduced by 85% with a 90% reduction in light at 23.5 °C compared to 71% with light reduction of 82% at 4 °C. At 4 °C and 1.792 $\mu E \cdot m^{-2} \cdot s^{-1}$ light intensity, or 1 °C and 0.32 $\mu E \cdot m^{-2} \cdot s^{-1}$, cell division was strongly inhibited. Another study on the interaction effects of light and temperature using Chlorella pyrenoidosa showed an increased saturation light intensity for algal growth at higher temperatures compared to lower temperatures. For higher temperatures, this alga exhibited a higher growth rate even at higher light intensities. The limitation on growth of algae under low temperature and high light intensity conditions is through effects on photosynthesis [210].

Converti et al. [80] studied the effects of temperature and nitrogen concentration on cell growth and lipid content in two strains of algae—Chlorella vulgaris and Nannochloropsis oculata. Reducing the nitrate concentrations in the growth media by 75% (1.5 to 0.375 g L^{-1} for Chlorella vulgaris and 0.3 to 0.075 L^{-1} for Nannochloropsis oculata), lipid accumulation tripled and doubled respectively, with only a small reduction in growth rate at optimal growth temperature. This result indicates that it may be possible to achieve higher lipid productivity for biofuels production by employing nitrogen limitation with fine temperature control. Interactions between salinity and nutrient concentrations (i.e., $NaNO_3$) for a salt-tolerant alga, Tetraselmis suecica, were also reported [144]. Optimum growth conditions were found at high salinity (25%–35%) and low nutrient concentrations (2, 4, and 8 mM $NaNO_3$). It was also reported that the total protein and protein content per cell increased with increase in salinity at constant nutrient concentration. Furthermore, a decrease in protein content was observed with increase in nitrate concentration at constant salinity. The transformation of nitrate to protein increased with the salinity and decreased with increasing nitrate concentrations.

High light (300 $\mu E \cdot m^{-2} \cdot s^{-1}$) with warm temperatures (22 °C) was observed to promote carbon fixation since this pathway acts as a sink for energy in conjunction with NO_3^- reduction pathways, which dissipate excess light energy. Confirming this hypothesis, a change in nitrogen source from NO_3^- to NH_4^+ controlling for light and temperature resulted in an increase of photorespiration, which is also a sink for excess energy. On the contrary, high light intensity (300 $\mu E \cdot m^{-2} \cdot s^{-1}$) and lower temperatures (12 °C) with NO_3^- as the nitrogen source resulted in increased energy dissipation by means of photorespiration and NO_3^- reduction (rather than via diversion to cellular carbohydrates). It was postulated that this result is due to lower carbon flows through the Calvin cycle at lower temperatures; in turn, this necessitates dissipation of the excess energy via photorespiration and NO_3^- reduction. Synergistic interaction effects of light, temperature, and nitrogen source on the transcription of five genes that are central to the carbon and nitrogen metabolism pathways in Thalassiosira pseudonana were also demonstrated [211].

The effects of environmental factors on biochemical composition of algae are summarized in Table 2. These results suggest that past growth conditions, interaction effects of various nutrients and environmental factors are all important to attain a thorough understanding of the behavior of algal cells in large-scale systems.

Table 2. Summary of general impact of environmental factors on biochemical composition of algae.

Factor	Organism	Conditions	Biochemical changes observed	References
Temperature	*Botryccoccus braunii*	Increased from 25 to 32 °C	Decrease in intracellular lipid content from 22% to 5% wt. Accumulation of polysaccharides	[66]
	Chlorella vulgaris	Increased from 20 to 38 °C	Decrease in starch resulting in increase in sucrose	[65]
		Increased from 10 to 38 °C	Transformation of L starch (high molecular weight) to S starch (low molecular weight) Reversible with temperature	[89]
	Haematococcus pluvialis	Increased from 20 to 30 °C	3-fold increase in astaxanthin formation	[91]
	Chlorococcum sp.	Temperature increase from 20 to 35 °C under nitrogen deprivation	Two fold increase in total carotenoid content	[92]
	Nitella mucronata Miquel	Increased from 5 to 20 °C	Increase in velocity of cytoplasmic streaming	[73]
Light	*Dunaliella viridis*	Darkness (No light)	Increase in total lipid content Decrease in free fatty acids, alcohol, sterol	[30]
	Nannochloropsis sp.	Light limited conditions	Increase in lipid content Increase in EPA * proportions	[26]
	Porphyridium cruentum	Red light	Enhanced Photosystem II relative to Photosystem I and phycobilisome	[212]
	Chlorella vulgaris	Red light	Increase in sucrose and starch formation	[113]
		Blue light	Increase in lipid fraction and alcohol-water insoluble non carbohydrate fraction	
pH	*Chlamydomonas acidophila*	pH 4.4	Denaturation of V-lysin	[128]
	Coccochloris peniocystis	pH decreased from 7.0 to 5.0 and 6.0	Decrease in total accumulated carbon and oxygen evolution	[130]

4. CONCLUSIONS

4.1. Algae as a Sustainable Biofuel Source

For the success of any sustainable biofuel, there are three principal considerations: technical feasibility; economic viability; and, resource sustainability. Algal-based biofuel is technically feasible. However, to date, economic viability has not been achieved. Furthermore, resource sustainability, in terms of land, water, nutrient and energy utilization, must be meticulously quantified for each type of production system in order for the feedstock to be considered truly "sustainable". With large-scale biofuel production processes, this water-energy-nutrient nexus is the subject of significant consideration and debate.

Sustainable use of wastewater and seawater resources poses one of the most significant challenges to large-scale production of algal-based biofuels. Algal biofuel production using waste water, brackish water, or sea water in open ponds or using closed photobioreactors, in which the evaporative water losses are negligible, are both potential solutions to this challenge.

Sustainable use of nutrients, such as nitrogen and phosphorous, also poses a serious challenge to large-scale production of algal-based biofuel. Energy intensive production methods for nitrogenous fertilizers and concerns about long-term availability of phosphorous accentuate these concerns. Phosphate is typically obtained by mining and some recent studies estimate that peak phosphorous could occur as early as 2030 [213]. Recycling nitrogen, phosphorous and other nutrients is a strategy to address some of these challenges while addressing other ecological issues such as eutrophication. Algae are capable of utilizing nutrients (including, nitrogen and phosphorus) from wastewater and thus could play a key role in nutrient recovery from waste waters [42,51,214,215, 216,217,218,219,220,221]. Maximizing algae production and minimizing costs associated with harvesting are critical to cost-effective nutrient removal system development [222].

4.2. Future Outlook

In this review, we discussed the effects of various nutritional factors (i.e., carbon, nitrogen, phosphorous, iron and trace metals) environmental factors (i.e., light, temperature and pH) and their interactions on algae growth and composition. During the last four decades, significant progress has been made toward a better understanding of algae growth. This includes an enhanced understanding of the effects of the nutrient availability and environmental factors on algae cell division and composition. However, interaction effects induced by multiple factors acting in concert have been generally underexplored. With advances in biotechnology and bioinformatics, a large volume of genetic information is now available rendering studies of interaction effects tractable. For example, full genomes for Chalmydomonas reinhardtii [223], Chlorella vulgaris [224] and Dunaleilla salina [225] have been sequenced. Many genetic manipulation tools have been developed for Chlamydomonas reinhardtii and others are being developed for other algal species. Continued research will not only lead to an enhanced understanding of basic algal cell biology but it will also aid in development of more accurate predictive models for algae growth [226]. Predictive models, in turn, can be used for development of automated optimal control systems for managing algae growth in large-scale production systems. This knowledge will be critical for successful scale-up of algae production systems for sustainable production of biofuels and other algae-based bioproducts.

CONFLICTS OF INTEREST

The authors declare no conflict of interest.

REFERENCES

1. Nichols, N.N.; Bothast, R.J. Production of Ethanol from Grain. In Genetic Improvement of Bioenergy Crops; Vermerris, W., Ed.; Springer: New York, NY, USA, 2008; pp. 75–88.

2. Goldemberg, J.; Coelho, S.T.; Guardabassi, P. The sustainability of ethanol production from sugarcane. Energy Policy 2008, 36, 2086–2097.

3. Samson, R.A.; Omielan, J.A. Switchgrass: A Potential Biomass Energy Crop for Ethanol Production. In Proceedings of the Thirteenth North American Prarie Conference, Windsor, ON, Canada, 6–9 August 1992; pp. 253–258.

4. Schmer, M.R.; Vogel, K.P.; Mitchell, R.B.; Perrin, R.K. Net energy of cellulosic ethanol from switchgrass. Proc. Natl. Acad. Sci. USA 2008, 105, 464–469.

5. Sheehan, J.; Aden, A.; Paustian, K.; Killian, K.; Brenner, J.; Walsh, M.; Nelson, R. Energy and environmental aspects of using corn stover for fuel ethanol. J. Ind. Ecol. 2003, 7, 117–146.

6. Ballesteros, I.; Negro, M.J.; Oliva, J.M.; Cabañas, A.; Manzanares, P.; Ballesteros, M. Ethanol production from steam-explosion pretreated wheat straw. Appl. Biochem. Biotechnol. 2006, 129–132, 496–508.

7. Kumar, D.; Murthy, G.S. Pretreatments and enzymatic hydrolysis of grass straws for ethanol production in the Pacific Northwest U.S. Biol. Eng. 2010, 3, 97–110.

8. Kumar, D.; Murthy, G.S. Impact of pretreatment and downstream processing technologies on economics and energy in cellulosic ethanol production. Biotechnol. Biofuels 2011, 4, 27.

9. Sanchez, O.J.; Cardona, C.A. Trends in biotechnological production of fuel ethanol from different feedstocks. Bioresour. Technol. 2008, 99, 5270–5295.

10. Gelfand, I.; Sahajpal, R.; Zhang, X.; Izaurralde, R.C.; Gross, K.L.; Robertson, G.P. Sustainable bioenergy production from marginal lands in the U.S. Midwest. Nature 2013, 493, 514–517.

11. Singh, J.; Gu, S. Commercialization potential of microalgae for biofuels production. Renew. Sustain. Energy Rev. 2010, 14, 2596–2610.

12. Amin, S. Review on biofuel oil and gas production processes from microalgae. Energy Convers. Manag. 2009, 50, 1834–1840.

13. Nakas, J.; Schaedle, M.; Parkinson, C.; Coonley, C.; Tanenbaum, S. System development for linked-fermentation production of solvents from algal biomass. Appl. Environ. Microbiol. 1983, 46, 1017–1023.

14. Sander, K.; Murthy, G.S. Life cycle analysis of algae biodiesel. Int. J. Life Cycle Assess. 2010, 15, 704–714.

15. Georgianna, D.R.; Mayfield, S.P. Exploiting diversity and synthetic biology for the production of algal biofuels. Nature 2012, 488, 329–335.

16. Sheehan, J. A Look Back at the U.S. Department of Energy's Aquatic Species Program: Biodiesel from Algae; National Renewable Energy Laboratory: Golden, CO, USA, 1998.

17. Ahmad, A.; Yasin, N.; Derek, C.; Lim, J. Microalgae as a sustainable energy source for biodiesel production: A review. Renew. Sustain. Energy Rev. 2011, 15, 584–593.

18. Deng, X.; Li, Y.; Fei, X. Microalgae: A promising feedstock for biodiesel. Afr. J. Microbiol. Res. 2009, 3, 1008–1014.

19. Khan, S.A.; Hussain, M.Z.; Prasad, S.; Banerjee, U. Prospects of biodiesel production from microalgae in India. Renew. Sustain. Energy Rev. 2009, 13, 2361–2372.

20. Mata, T.M.; Martins, A.A.; Caetano, N.S. Microalgae for biodiesel production and other applications: A review. Renew. Sustain. Energy Rev. 2010, 14, 217–232.

21. Eshaq, F.S.; Ali, M.N.; Mohd, M.K. Spirogyra biomass a renewable source for biofuel (bioethanol) production. Int. J. Eng. Sci. Technol. 2010, 2, 7045–7054.

22. Harun, R.; Danquah, M.K.; Forde, G.M. Microalgal biomass as a fermentation feedstock for bioethanol production. J. Chem. Technol. Biotechnol. 2010, 85, 199–203.

23. John, R.P.; Anisha, G.; Nampoothiri, K.M.; Pandey, A. Micro and macroalgal biomass: A renewable source for bioethanol. Bioresour. Technol. 2011, 102, 186–193.

24. Moen, E. Biological Degradation of Brown Seaweeds. Ph.D. Thesis, Norwegian University of Science and Technology, Trondheim, Norway, November 1997.

25. Morris, I.; Glover, H.; Yentsch, C. Products of photosynthesis by marine phytoplankton: The effect of environmental factors on the relative rates of protein synthesis. Mar. Biol. 1974, 27, 1–9.

26. Sukenik, A.; Carmeli, Y.; Berner, T. Regulation of fatty acid composition by irradiance level in the eustigmatophyte Nannochloropsis sp. J. Phycol. 1989, 25, 686–692.

27. Thompson, P.A.; Guo, M.; Harrison, P.J. Effects of variation in temperature. I. On the biochemical composition of eight species of marine phytoplankton. J. Phycol. 1992, 28, 481–488.

28. Chen, C.Y.; Durbin, E.G. Effects of pH on the growth and carbon uptake of marine phytoplankton. Mar. Ecol.-Prog. Ser. 1994, 109, 83–94.

29. Kilham, S.; Kreeger, D.; Goulden, C.; Lynn, S. Effects of nutrient limitation on biochemical constituents of Ankistrodesmus falcatus. Freshw. Biol. 1997, 38, 591–596.

30. Gordillo, F.J.L.; Goutx, M.; Figueroa, F.L.; Niell, F.X. Effects of light intensity, CO2 and nitrogen supply on lipid class composition of Dunaliella viridis. J. Appl. Phycol. 1998, 10, 135–144.

31. Visviki, I.; Santikul, D. The pH tolerance of Chlamydomonas applanata (Volvocales, Chlorophyta). Arch. Environ. Contam. Toxicol. 2000, 38, 147–151.

32. Brennan, L.; Owende, P. Biofuels from microalgae—A review of technologies for production, processing, and extractions of biofuels and co-products. Renew. Sustain. Energy Rev. 2010, 14, 557–577.

33. Potter, A.L.; Hassid, W.Z. Starch. II. Molecular weights of amyloses and amylopectins from starches of various plant origins. J. Am. Chem. Soc. 1948, 70, 3774–3777.

34. Roessler, P.G. Environmental control of glycerolipid metabolism in microalgae: Commercial implications and future research directions. J. Phycol. 1990, 26, 393–399.

35. Sharma, K.K.; Schuhmann, H.; Schenk, P.M. High lipid induction in microalgae for biodiesel production. Energies 2012, 5, 1532–1553.

36. Chen, C.Y.; Yeh, K.L.; Aisyah, R.; Lee, D.J.; Chang, J.S. Cultivation, photobioreactor design and harvesting of microalgae for biodiesel production: A critical review. Bioresour. Technol. 2011, 102, 71–81.

37. Du, Z.; Li, Y.; Wang, X.; Wan, Y.; Chen, Q.; Wang, C.; Lin, X.; Liu, Y.; Chen, P.; Ruan, R. Microwave-assisted pyrolysis of microalgae for biofuel production. Bioresour. Technol. 2011, 102, 4890–4896.

38. Zou, S.P.; Wu, Y.L.; Yang, M.D.; Li, C.; Tong, J.M. Thermochemical catalytic liquefaction of the marine microalgae Dunaliella tertiolecta and characterization of bio-oils. Energy Fuels 2009, 23, 3753–3758.

39. Ueno, Y.; Kurano, N.; Miyachi, S. Ethanol production by dark fermentation in the marine green alga, Chlorococcum littorale. J. Ferment. Bioeng. 1998, 86, 38–43.

40. Sharma, Y.C.; Singh, B. Development of biodiesel: Current scenario. Renew. Sustain. Energy Rev. 2009, 13, 1646–1651.

41. Goyal, H.B.; Seal, D.; Saxena, R.C. Bio-fuels from thermochemical conversion of renewable resources: A review. Renew. Sustain. Energy Rev. 2008, 12, 504–517.

42. Cantrell, K.B.; Ducey, T.; Ro, K.S.; Hunt, P.G. Livestock waste-to-bioenergy generation opportunities. Bioresour. Technol. 2008, 99, 7941–7953.

43. Miao, X.; Wu, Q. High yield bio-oil production from fast pyrolysis by metabolic controlling of Chlorella protothecoides. J. Biotechnol. 2004, 110, 85–93.

44. Duan, P.; Jin, B.; Xu, Y.; Yang, Y.; Bai, X.; Wang, F.; Zhang, L.; Miao, J. Thermo-chemical conversion of Chlorella pyrenoidosa to liquid biofuels. Bioresour. Technol. 2013, 133, 197–205.

45. Chisti, Y. Constraints to commercialization of algal fuels. J. Biotechnol. 2013, 167, 201–214.

46. Pragya, N.; Pandey, K.K.; Sahoo, P. A review on harvesting, oil extraction and biofuels production technologies from microalgae. Renew. Sustain. Energy Rev. 2013, 24, 159–171.

47. Suali, E.; Sarbatly, R. Conversion of microalgae to biofuel. Renew. Sustain. Energy Rev. 2012, 16, 4316–4342.

48. Becker, E.W. Micro-algae as a source of protein. Biotechnol. Adv. 2007, 25, 207–210.

49. Borowitzka, L.J.; Borowitzka, M.A. Commercial production of-carotene by Dunaliella salina in open ponds. Bull. Mar. Sci. 1990, 47, 244–252.

50. Chen, F.; Zhang, Y. High cell density mixotrophic culture of Spirulina platensis on glucose for phycocyanin production using a fed-batch system. Enzyme Microb. Technol. 1997, 20, 221–224.

51. Hu, Q.; Sommerfeld, M.; Jarvis, E.; Ghirardi, M.; Posewitz, M.; Seibert, M.; Darzins, A. Microalgal triacylglycerols as feedstocks for biofuel production: Perspectives and advances. Plant J. 2008, 54, 621–639.

52. Chisti, Y. Biodiesel from microalgae. Biotechnol. Adv. 2007, 25, 294–306.

53. Beer, L.L.; Boyd, E.S.; Peters, J.W.; Posewitz, M.C. Engineering algae for biohydrogen and biofuel production. Curr. Opin. Biotechnol. 2009, 20, 264–271.

54. Benemann, J. Feasibility analysis of photobiological hydrogen production. Int. J. Hydrog. Energy 1997, 22, 979–987.

55. Hempel, F.; Bozarth, A.S.; Lindenkamp, N.; Klingl, A.; Zauner, S.; Linne, U.; Steinbüchel, A.; Maier, U.G. Microalgae as bioreactors for bioplastic production. Microb. Cell Fact. 2011, 10, 81.

56. Jeong, M.L.; Gillis, J.M.; Hwang, J.Y. Carbon dioxide mitigation by microalgal photosynthesis. Bull. Korean Chem. Soc. 2003, 24, 1763–1766.

57. Keffer, J.E.; Kleinheinz, G.T. Use of Chlorella vulgaris for CO2 mitigation in a photobioreactor. J. Ind. Microbiol. Biotechnol. 2002, 29, 275–280.

58. Hu, Q. Environmental Effects on Cell Composition. In Handbook of Microalgal Culture: Biotechnology and Applied Phycology; Richmond, A., Ed.; Blackwell: Oxford, UK, 2004; pp. 83–93.

59. Hirano, A.; Ueda, R.; Hirayama, S.; Ogushi, Y. CO2 fixation and ethanol production with microalgal photosynthesis and intracellular anaerobic fermentation. Energy 1997, 22, 137–142.

60. Nguyen, M.T.; Choi, S.P.; Lee, J.; Lee, J.H.; Sim, S.J. Hydrothermal acid pretreatment of Chlamydomonas reinhardtii biomass for ethanol production. J. Microbiol. Biotechnol. 2009, 19, 161–166.

61. Spolaore, P.; Joannis-Cassan, C.; Duran, E.; Isambert, A. Commercial applications of microalgae. J. Biosci. Bioeng. 2006, 101, 87–96.

62. Hasan, M.R.; Chakrabarti, R. Use of Algae and Aquatic Macrophytes as Feed in Small-Scale Aquaculture: A Review; Food and Agriculture Organization of the United Nations (FAO): Rome, Italy, 2009.

63. Zepka, L.Q.; Jacob-Lopes, E.; Goldbeck, R.; Souza-Soares, L.A.; Queiroz, M.I. Nutritional evaluation of single-cell protein produced by Aphanothece microscopica Nägeli. Bioresour. Technol. 2010, 101, 7107–7111.

64. Walker, J.B. Inorganic micronutrient requirements of Chlorella. II. Quantitative requirements for iron, manganese, and zinc. Arch. Biochem. Biophys. 1954, 53, 1–8.

65. Nakamura, Y.; Miyachi, S. Effect of temperature on starch degradation in Chlorella vulgaris 11 h cells. Plant Cell Physiol. 1982, 23, 333–341.

66. Kalacheva, G.; Zhila, N.; Volova, T.; Gladyshev, M. The effect of temperature on the lipid composition of the green alga Botryococcus. Microbiology 2002, 71, 286–293.

67. Fábregas, J.; Maseda, A.; Domínguez, A.; Otero, A. The cell composition of Nannochloropsis sp. changes under different irradiances in semicontinuous culture. World J. Microbiol. Biotechnol. 2004, 20, 31–35.

68. Renaud, S.M.; Thinh, L.V.; Lambrinidis, G.; Parry, D.L. Effect of temperature on growth, chemical composition and fatty acid composition of tropical Australian microalgae grown in batch cultures. Aquaculture 2002, 211, 195–214.

69. Rhee, G.Y. Effects of environmental factors and their interactions on phytoplankton growth. Adv. Microb. Ecol. 1982, 6, 33–74.

70. Harris, G.P. Phytoplankton Ecology: Structure, Function and Fluctuation; Chapman and Hall: New York, NY, USA, 1986.

71. Darley, W.M. Algal Biology: A Physiological Approach; Blackwell: Oxford, UK, 1982; Volume 9, pp. 1–168.

72. Hope, A.B.; Walker, N.A. The Physiology of Giant Algal Cells; Cambridge University Press: London, UK, 1975.

73. Raven, J.A.; Geider, R.J. Temperature and algal growth. New Phytol. 1988, 110, 441–461.

74. Vonshak, A.; Torzillo, G. Environmental Stress Physiology. In Handbook of Microalgal Culture; Richmond, A., Ed.; Blackwell: Oxford, UK, 2004; pp. 57–82.

75. Thompson, G.A., Jr. Lipids and membrane function in green algae. Biochim. Biophys. Acta 1996, 1302, 17–45.

76. Harwood, J.L. Involvement of Chloroplast Lipids in the Reaction of Plants Submitted to Stress. In Lipids in Photosynthesis: Structure, Function and Genetics; Siegenthaler, P.A., Murata, N., Eds.; Springer: Berlin, Germany, 2004; Volume 6, pp. 287–302.

77. Guschina, I.A.; Harwood, J.L. Algal Lipids and Effect of the Environment on Their Biochemistry. In Lipids in Aquatic Ecosystems; Springer: Berlin, Germany, 2009; pp. 1–24.

78. Lynch, D.V.; Thompson, G.A., Jr. Low temperature-induced alterations in the chloroplast and microsomal membranes of Dunaliella salina. Plant Physiol. 1982, 69, 1369–1375.

79. Nishida, I.; Murata, N. Chilling sensitivity in plants and cyanobacteria: The crucial contribution of membrane lipids. Annu. Rev. Plant Biol. 1996, 47, 541–568.

80. Converti, A.; Casazza, A.A.; Ortiz, E.Y.; Perego, P.; del Borghi, M. Effect of temperature and nitrogen concentration on the growth and lipid content of Nannochloropsis oculata and Chlorella vulgaris for biodiesel production. Chem. Eng. Process. 2009, 48, 1146–1151.

81. Konopka, A.; Brock, T.D. Effect of temperature on blue-green algae (cyanobacteria) in Lake Mendota. Appl. Environ. Microbiol. 1978, 36, 572–576.

82. Cuhel, R.L.; Ortner, P.B.; Lean, D.R.S. Night synthesis of protein by algae. Limnol. Oceanogr. 1984, 29, 731–744.

83. Rhee, G.Y.; Gotham, I.J. The effect of environmental factors on phytoplankton growth: Temperature and the interactions of temperature with nutrient limitation. Limnol. Oceanogr. 1981, 26, 635–648.

84. Kakinuma, M.; Coury, D.; Kuno, Y.; Itoh, S.; Kozawa, Y.; Inagaki, E.; Yoshiura, Y.; Amano, H. Physiological and biochemical responses to thermal and salinity stresses in a sterile mutant of Ulva pertusa (Ulvales, Chlorophyta). Mar. Biol. 2006, 149, 97–106.

85. Emerson, R.; Stauffer, J.; Umbreit, W. Relationships between phosphorylation and photosynthesis in Chlorella. Am. J. Bot. 1944, 31, 107–120.

86. Mitsui, A.; United States-Japan Cooperative Science Program; National Science Foundation (U.S.). Biological Solar Energy Conversion; Academic Press: New York, NY, USA, 1977.

87. Nakamura, Y. Change in molecular weight distribution in starch when degraded at different temperatures in Chlorella vulgaris. Plant Sci. Lett. 1983, 30, 259–265.

88. Nakamura, Y.; Miyachi, S. Change in starch photosynthesized at different temperatures in Chlorella. Plant Sci. Lett. 1982, 27, 1–6.

89. Nakamura, Y.; Imamura, M. Change in properties of starch when photosynthesized at different temperatures in Chlorella vulgaris. Plant Sci. Lett. 1983, 31, 123–131.

90. Armstrong, G.A.; Hearst, J.E. Carotenoids 2: Genetics and molecular biology of carotenoid pigment biosynthesis. FASEB J. 1996, 10, 228–237.

91. Tjahjono, A.E.; Hayama, Y.; Kakizono, T.; Terada, Y.; Nishio, N.; Nagai, S. Hyper-accumulation of astaxanthin in a green alga Haematococcus pluvialis at elevated temperatures. Biotechnol. Lett. 1994, 16, 133–138.

92. Liu, B.H.; Lee, Y.K. Secondary carotenoids formation by the green alga Chlorococcum sp. J. Appl. Phycol. 2000, 12, 301–307.

93. Tripathi, U.; Sarada, R.; Ravishankar, G. Effect of culture conditions on growth of green alga—Haematococcus pluvialis and astaxanthin production. Acta Physiol. Plant. 2002, 24, 323–329.

94. NREL Web Page. Dynamic Maps, GIS Data, and Analysis Tools—Solar Maps. Available online: http://www.nrel.gov/gis/solar.html (accessed on 9 May 2013).

95. Stockenreiter, M.; Haupt, F.; Graber, A.K.; Seppälä, J.; Spilling, K.; Tamminen, T.; Stibor, H. Functional group richness: Implications of biodiversity for light use and lipid yield in microalgae. J. Phycol. 2013. in press.

96. Sorokin, C.; Krauss, R.W. The Effects of light intensity on the growth rates of green algae. Plant Physiol. 1958, 33, 109–113.

97. Dubinsky, Z.; Matsukawa, R.; Karube, I. Photobiological aspects of algal mass culture. J. Mar. Biotechnol. 1995, 2, 61–65.

98. Berner, T.; Dubinsky, Z.; Wyman, K.; Falkowski, P.G. Photoadaptation and the "package" effect in Dunaliella tertiolecta (chlorophycae). J. Phycol. 1989, 25, 70–78.

99. Mock, T.; Kroon, B.M.A. Photosynthetic energy conversion under extreme conditions—II: The significance of lipids under light limited growth in Antarctic sea ice diatoms. Phytochemistry 2002, 61, 53–60.

100. You, T.; Barnett, S.M. Effect of light quality on production of extracellular polysaccharides and growth rate of Porphyridium cruentum. Biochem. Eng. J. 2004, 19, 251–258.

101. Brody, M.; Vatter, A.E. Observations on cellular structures of Porphyridium cruentum. J. Biophys. Biochem. Cytol. 1959, 5, 289–294.

102. Iqbal, M.; Zafar, S. Effects of photon flux density, CO2, aeration rate, and inoculum density on growth and extracellular polysaccharide production by Porphyridium cruentum. Folia Microbiol. 1993, 38, 509–514.

103. Smith, R.; Cavaletto, J.; Eadie, B.; Gardner, W. Growth and lipid composition of high Arctic ice algae during the spring bloom at Resolute, Northwest Territories, Canada. Mar. Ecol. Prog. Ser. 1993, 97, 19–29.

104. Cohen, Z. Porphyridium Cruentum. In Chemicals from Microalgae; Cohen, Z., Ed.; CRC Press: Boca Raton, FL, USA, 1999; pp. 1–24.

105. Renaud, S.; Parry, D.; Thinh, L.V.; Kuo, C.; Padovan, A.; Sammy, N. Effect of light intensity on the proximate biochemical and fatty acid composition of Isochrysis sp. and Nannochloropsis oculata for use in tropical aquaculture. J. Appl. Phycol. 1991, 3, 43–53.

106. Orcutt, D.M.; Patterson, G.W. Effect of light intensity upon lipid composition of Nitzschia closterium (Cylindrotheca fusiformis). Lipids 1974, 9, 1000–1003.

107. Molina Grima, E.; Garcia Camacho, F.; Acien Fernandez, F. Production of EPA from Phaeodactylum Tricornumtum. In Chemicals from Microalgae; Cohen, Z., Ed.; CRC Press: Boca Raton, FL, USA, 1999; pp. 57–92.

108. Wu, X.; Merchuk, J.C. A model integrating fluid dynamics in photosynthesis and photoinhibition processes. Chem. Eng. Sci. 2001, 56, 3527–3538.

109. Long, S.; Humphries, S.; Falkowski, P.G. Photoinhibition of photosynthesis in nature. Annu. Rev. Plant Biol. 1994, 45, 633–662.

110. Voigt, J.; Münzner, P. Blue light-induced lethality of a cell wall-deficient mutant of the unicellular green alga Chlamydomonas reinhardtii. Plant Cell Physiol. 1994, 35, 99–106.

111. Borodin, V.B. Effect of red and blue light on acclimation of Chlamydomonas reinhardtii to CO2-limiting conditions. Rus. J. Plant Physiol. 2008, 55, 441–448.

112. Emerson, R.; Lewis, C.M. The dependence of the quantum yield of Chlorella photosynthesis on wave lenghth of light. Am. J. Bot. 1943, 30, 165–178.

113. Miyachi, S.; Kamiya, A. Wavelength effects on photosynthetic carbon metabolism in Chlorella. Plant Cell Physiol. 1978, 19, 277–288.

114. Fernanda Pessoa, M. Harmful effects of UV radiation in algae and aquatic macrophytes—A review. Emir. J. Food Agric. 2012, 24, 510–526.

115. Xue, L.; Zhang, Y.; Zhang, T.; An, L.; Wang, X. Effects of enhanced ultraviolet-B radiation on algae and cyanobacteria. Crit. Revi. Microbiol. 2005, 31, 79–89.

116. Rastogi, R.P.; Incharoensadki, A. UV radiation-induced accumulation of photoprotective compounds in the green alga Tetraspora sp. CU2551. Plant Physiol. Biochem. 2013, 70, 7–13.

117. Goldman, J.C. Letter: Carbon dioxide and pH: Effect on species succession of algae. Science 1973, 182, 306–307.

118. Hansen, P.J. Effect of high pH on the growth and survival of marine phytoplankton: Implications for species succession. Aquat. Microb. Ecol. 2002, 28, 279–288.

119. Goldman, J.C.; Azov, Y.; Riley, C.B.; Dennett, M.R. The effect of pH in intensive microalgal cultures. I. biomass regulation. J. Exp. Mar. Biol. Ecol. 1982, 57, 1–13.

120. Pruder, G.D.; Bolton, E.T. The role of CO2 enrichment of aerating gas in the growth of an estuarine diatom. Aquaculture 1979, 17, 1–15.

121. Azov, Y. Effect of pH on inorganic carbon uptake in algal cultures. Appl. Environ. Microbiol. 1982, 43, 1300–1306.

122. Nielsen, E.S. Marine Photosynthesis: With Special Emphasis on the Ecological Aspects; Elsevier: Amsterdam, The Netherlands, 1975; Volume 13.

123. Rotatore, C.; Colman, B. The acquisition and accumulation of inorganic carbon by the unicellular green alga Chlorella ellipsoidea. Plant Cell Environ. 1991, 14, 377–382.

124. Guckert, J.B.; Cooksey, K.E. Triglyceride accumulation and fatty acid profile changes in Chlorella (Chlorophyta) during high pH induced cell cycle inhibition. J. Phycol. 1990, 26, 72–79.

125. Gensemer, R.W.; Smith, R.E.H.; Duthie, H.C. Comparative effects of pH and aluminum on silica limited growth and nutrient uptake in Asterionella ralfsii var. Americana (Bacillariophyceae). J. Phycol. 1993, 29, 36–44.

126. Sunda, W. The Relationship between Cupric Ion Activity and the Toxicity of Copper to Phytoplankton. Ph.D. Thesis, Massachusetts Institute of Technology, Cambridge, MA, USA, April 1975.

127. Anderson, D.M.; Morel, F.M.M. Copper sensitivity of Gonyaulax tamarensis. Limnol. Oceanogr. 1978, 23, 283–295.

128. Visviki, I.; Palladino, J. Growth and cytology of Chlamydomonas acidophila under acidic stress. Bull. Environ. Contam. Toxicol. 2001, 66, 623–630.

129. Hargreaves, J.; Whitton, B. Effect of pH on growth of acid stream algae. Eur. J. Phycol. 1976, 11, 215–223.

130. Coleman, J.R.; Colman, B. Inorganic carbon accumulation and photosynthesis in a blue-green alga as a function of external pH. Plant Physiol. 1981, 67, 917–921.

131. Terry, N.; Abadía, J. Function of iron in chloroplasts. J. Plant Nutr. 1986, 9, 609–646.

132. Gehl, K.A.; Colman, B. Effect of external pH on the internal pH of Chlorella saccharophila. Plant Physiol. 1985, 77, 917–921.

133. Lane, A.E.; Burris, J.E. Effects of environmental pH on the internal pH of Chlorella pyrenoidosa, Scenedesmus quadricauda, and Euglena mutabilis. Plant Physiol. 1981, 68, 439–442.

134. Fuggi, A.; Pinto, G.; Pollio, A.; Taddei, R. The role of glycerol in osmoregulation of the acidophilic alga Dunaliella acidophila (Volvocales, Chlorophyta): Effect of solute stress on photosynthesis, respiration and glycerol synthesis. Phycologia 1988, 27, 439–446.

135. Tatsuzawa, H.; Takizawa, E.; Wada, M.; Yamamoto, Y. Fatty acid and lipid composition of the acidophilic green alga Chlamydomonas sp. 1. J. Phycol. 1996, 32, 598–601.

136. Poerschmann, J.; Spijkerman, E.; Langer, U. Fatty acid patterns in Chlamydomonas sp. as a marker for nutritional regimes and temperature under extremely acidic conditions. Microb. Ecol. 2004, 48, 78–89.

137. Moazami-Goudarzi, M.; Colman, B. Changes in carbon uptake mechanisms in two green marine algae by reduced seawater pH. J. Exp. Mar. Biol. Ecol. 2012, 413, 94–99.

138. Moroney, J.V.; Tolbert, N.E. Inorganic carbon uptake by Chlamydomonas reinhardtii. Plant Physiol. 1985, 77, 253–258.

139. Colman, B.; Huertas, I.E.; Bhatti, S.; Dason, J.S. The diversity of inorganic carbon acquisition mechanisms in eukaryotic microalgae. Funct. Plant Biol. 2002, 29, 261–270.

140. Ramanan, R.; Vinayagamoorthy, N.; Sivanesan, S.D.; Kannan, K.; Chakrabarti, T. Influence of CO2 concentration on carbon concentrating mechanisms in cyanobacteria and green algae: A proteomic approach. Algae 2012, 27, 295–301.

141. Zhila, N.O.; Kalacheva, G.S.; Volova, T.G. Effect of salinity on the biochemical composition of the alga Botryococcus braunii Kütz IPPAS H-252. J. Appl. Phycol. 2011, 23, 47–52.

142. Renaud, S.; Parry, D. Microalgae for use in tropical aquaculture II: Effect of salinity on growth, gross chemical composition and fatty acid composition of three species of marine microalgae. J. Appl. Phycol. 1994, 6, 347–356.

143. Ben-Amotz, A.; Tornabene, T.G.; Thomas, W.H. Chemical profile of selected species of microalgae with emphasis on lipids. J. Phycol. 1985, 21, 72–81.

144. Fabregas, J.; Abalde, J.; Herrero, C.; Cabezas, B.; Veiga, M. Growth of the marine microalga Tetraselmis suecica in batch cultures with different salinities and nutrient concentrations. Aquaculture 1984, 42, 207–215.

145. Xu, X.Q.; Beardall, J. Effect of salinity on fatty acid composition of a green microalga from an antarctic hypersaline lake. Phytochemistry 1997, 45, 655–658.

146. Takagi, M.; Yoshida, T. Effect of salt concentration on intracellular accumulation of lipids and triacylglyceride in marine microalgae Dunaliella cells. J. Biosci. Bioeng. 2006, 101, 223–226.

147. Rao, A.R.; Dayananda, C.; Sarada, R.; Shamala, T.; Ravishankar, G. Effect of salinity on growth of green alga Botryococcus braunii and its constituents. Bioresour. Technol. 2007, 98, 560–564.

148. Vazquez-Duhalt, R.; Arredondo-Vega, B.O. Haloadaptation of the green alga Botryococcus braunii (race A). Phytochemistry 1991, 30, 2919–2925.

149. Titman, D. Ecological competition between algae: Experimental confirmation of resource-based competition theory. Science 1976, 192, 463–465.

150. Berman-Frank, I.; Dubinsky, Z. Balanced growth and aquatic plants: Myth or reality? Phytoplankton use the imbalance between carbon assimilation and biomass production to their strategic advantage. Bioscience 1999, 49, 29–37.

151. Moss, B. The influence of environmental factors on the distribution of freshwater algae: An experimental study: II. The role of pH and the carbon dioxide-bicarbonate system. J. Ecol. 1973, 61, 157–177.

152. Riebesell, U.; Revill, A.T.; Holdsworth, D.G.; Volkman, J.K. The effects of varying CO2 concentration on lipid composition and carbon isotope fractionation in Emiliania huxleyi. Geochim. Cosmochim. Acta 2000, 64, 4179–4192.

153. Muradyan, E.; Klyachko-Gurvich, G.; Tsoglin, L.; Sergeyenko, T.; Pronina, N. Changes in lipid metabolism during adaptation of the Dunaliella salina photosynthetic apparatus to high CO2 concentration. Rus. J. Plant Physiol. 2004, 51, 53–62.

154. Tsuzuki, M.; Ohnuma, E.; Sato, N.; Takaku, T.; Kawaguchi, A. Effects of CO2 concentration during growth on fatty acid composition in microalgae. Plant Physiol. 1990, 93, 851–856.

155. Gordillo, F.J.L.; Jiménez, C.; Figueroa, F.L.; Niell, F.X. Effects of increased atmospheric CO2 and N supply on photosynthesis,

growth and cell composition of the cyanobacterium Spirulina platensis (Arthrospira). J. Appl. Phycol. 1998, 10, 461–469.

156. Fujita, R.M.; Wheeler, P.A.; Edwards, R.L. Metabolic regulation of ammonia uptake by Ulva rigida (Chlorophyta): A compartmental analysis of the rate limiting step for uptake. J. Phycol. 1988, 24, 560–566.

157. Vergara, J.; Bird, K.; Niell, F. Nitrogen assimilation following NH4+ pulses in the red alga Gracilariopsis lemaneiformis: Effect on C metabolism. Mar. Ecol. Prog. Ser. 1995, 122, 253–263.

158. Shifrin, N.S.; Chisholm, S.W. Phytoplankton lipids: Interspecific differences and effects of nitrate, silicate and light-dark cycles. J. Phycol. 1981, 17, 374–384.

159. Wang, Z.T.; Ullrich, N.; Joo, S.; Waffenschmidt, S.; Goodenough, U. Algal lipid bodies: Stress induction, purification, and biochemical characterization in wild-type and starchless Chlamydomonas reinhardtii. Eukaryot. Cell 2009, 8, 1856–1868.

160. Demirbas, A. Use of algae as biofuel sources. Energy Convers. Manag. 2010, 51, 2738–2749.

161. Takagi, M.; Watanabe, K.; Yamaberi, K.; Yoshida, T. Limited feeding of potassium nitrate for intracellular lipid and triglyceride accumulation of Nannochloris sp. UTEX LB1999. Appl. Microbiol. Biotechnol. 2000, 54, 112–117.

162. Stephenson, A.L.; Dennis, J.S.; Howe, C.J.; Scott, S.A.; Smith, A.G. Influence of nitrogen-limitation regime on the production by Chlorella vulgaris of lipids for biodiesel feedstocks. Biofuels 2010, 1, 47–58.

163. Fogg, G. Photosynthesis and formation of fats in a diatom. Ann. Bot. 1956, 20, 265–285.

164. Holm-Hansen, O.; Nishida, K.; Moses, V.; Calvin, M. Effects of mineral salts on short-term incorporation of carbon dioxide in Chlorella. J. Exp. Bot. 1959, 10, 109–124.

165. Lynn, S.G.; Kilham, S.S.; Kreeger, D.A.; Interlandi, S.J. Effect of nutrient availability on the biochemical and elemental stoichiometry in the freshwater diatom Stephanodiscus minutulus (Bacillariophyceae). J. Phycol. 2000, 36, 510–522.

166. Heraud, P.; Wood, B.R.; Tobin, M.J.; Beardall, J.; McNaughton, D. Mapping of nutrient-induced biochemical changes in living algal cells using synchrotron infrared microspectroscopy. FEMS Microbiol. Lett. 2005, 249, 219–225.

167. Li, Y.; Horsman, M.; Wang, B.; Wu, N.; Lan, C.Q. Effects of nitrogen sources on cell growth and lipid accumulation of green alga Neochloris oleoabundans. Appl. Microbiol. Biotechnol. 2008, 81, 629–636.

168. Metting, F.B., Jr. Biodiversity and application of microalgae. J. Ind. Microbiol. 1996, 17, 477–489.

169. Kolber, Z.; Zehr, J.; Falkowski, P. Effects of growth irradiance and nitrogen limitation on photosynthetic energy conversion in photosystem II. Plant Physiol. 1988, 88, 923–929.

170. Richardson, B.; Orcutt, D.; Schwertner, H.; Martinez, C.L.; Wickline, H.E. Effects of nitrogen limitation on the growth and composition of unicellular algae in continuous culture. Appl. Microbiol. 1969, 18, 245–250.

171. Collier, J.L.; Grossman, A. Chlorosis induced by nutrient deprivation in Synechococcus sp. strain PCC 7942: Not all bleaching is the same. J. Bacteriol. 1992, 174, 4718–4726.

172. Round, F.E. The Ecology of Algae; Cambridge University Press: Cambridge, UK, 1984.

173. Ben-Amotz, A.; Avron, M. The biotechnology of cultivating the halotolerant alga Dunaliella. Trends Biotechnol. 1990, 8, 121–126.

174. Borowitzka, M.A.; Huisman, J.M.; Osborn, A. Culture of the astaxanthin-producing green alga Haematococcus pluvialis. Effects of nutrients on growth and cell type. J. Appl. Phycol. 1991, 3, 295–304.

175. Harker, M.; Tsavalos, A.J.; Young, A.J. Factors responsible for astaxanthin formation in the chlorophyte Haematococcus pluvialis. Bioresour. Technol. 1996, 55, 207–214.

176. Zhekisheva, M.; Boussiba, S.; Khozin Goldberg, I.; Zarka, A.; Cohen, Z. Accumulation of oleic acid in Haematococcus pluvialis (Chlorophyceae) under nitrogen starvation or high ligh is

correlated with that of astaxanthin esters. J. Phycol. 2002, 38, 325–331.

177. Larned, S. Nitrogen-versus phosphorus-limited growth and sources of nutrients for coral reef macroalgae. Mar. Biol. 1998, 132, 409–421.

178. Borchardt, J.A.; Azad, H.S. Biological extraction of nutrients. J. Water Pollut. Control Fed. 1968, 40, 1739–1754.

179. Barsanti, L.; Gualtieri, P. Algae: Anatomy, Biochemistry, and Biotechnology, 1st ed.; CRC Press: Boca Raton, FL, USA, 2005.

180. Li, X.; Hu, H.Y.; Gan, K.; Sun, Y.X. Effects of different nitrogen and phosphorus concentrations on the growth, nutrient uptake, and lipid accumulation of a freshwater microalga Scenedesmus sp. Bioresour. Technol. 2010, 101, 5494–5500.

181. Sato, N.; Hagio, M.; Wada, H.; Tsuzuki, M. Environmental effects on acidic lipids of thylakoid membranes. Biochem. Soc. Trans. 2000, 28, 912–914.

182. Reitan, K.I.; Rainuzzo, J.R.; Olsen, Y. Effect of nutrient limitation on fatty acid and lipid content of marine microalgae. J. Phycol. 1994, 30, 972–979.

183. Healey, F.P. Phosphate. Biol. Cyanobacteria 1982, 19, 105–124.

184. Healey, F.P.; Hendzel, L.L. Indicators of phosphorus and nitrogen deficiency in five algae in culture. J. Fish. Board Can. 1979, 36, 1364–1369.

185. Kobayashi, M.; Kakizono, T.; Nagai, S. Enhanced carotenoid biosynthesis by oxidative stress in acetate-induced cyst cells of a green unicellular alga, Haematococcus pluvialis. Appl. Environ. Microbiol. 1993, 59, 867–873.

186. Theodorou, M.E.; Elrifi, I.R.; Turpin, D.H.; Plaxton, W.C. Effects of phosphorus limitation on respiratory metabolism in the green alga Selenastrum minutum. Plant Physiol. 1991, 95, 1089–1095.

187. Bruland, K.W.; Donat, J.R.; Hutchins, D.A. Interactive influences of bioactive trace metals on biological production in oceanic waters. Limnol. Oceanogr. 1991, 36, 1555–1577.

188. Parent, L.; Twiss, M.R.; Campbell, P.G.C. Influences of natural dissolved organic matter on the interaction of aluminum with the microalga Chlorella: A test of the free-ion model of trace metal toxicity. Environ. Sci. Technol. 1996, 30, 1713–1720.

189. McKay, R.M.L.; La Roche, J.; Yakunin, A.F.; Durnford, D.G.; Geider, R.J. Accumulation of ferredoxin and flavodoxin in a marine diatom in response to Fe. J. Phycol. 1999, 35, 510–519.

190. Sandmann, G.; Malkin, R. Iron-sulfur centers and activities of the photosynthetic electron transport chain in iron-deficient cultures of the blue-green alga Aphanocapsa. Plant Physiol. 1983, 73, 724–728.

191. Roche, J.; Geider, R.J.; Graziano, L.M.; Murray, H.; Lewis, K. Induction of specific proteins in eukaryotic algae grown under iron-, phosphorus-, or nitrogen-deficient conditions. J. Phycol. 1993, 29, 767–777.

192. Straus, N.A. Iron Deprivation: Physiology and Gene Regulation. In The Molecular Biology of Cyanobacteria; Bryant, D.A., Ed.; Springer: Berlin, Germany, 2004; Volume 1, pp. 731–750.

193. Raven, J.A. Energetics and Transport in Aquatic Plants; Alan R. Liss: New York, NY, USA, 1984.

194. Greene, R.M.; Geider, R.J.; Kolber, Z.; Falkowski, P.G. Iron-induced changes in light harvesting and photochemical energy conversion processes in eukaryotic marine algae. Plant Physiol. 1992, 100, 565–575.

195. Liu, Z.Y.; Wang, G.C.; Zhou, B.C. Effect of iron on growth and lipid accumulation in Chlorella vulgaris. Bioresour. Technol. 2008, 99, 4717–4722.

196. Van Leeuwe, M.A.; Stefels, J. Effects of iron and light stress on the biochemical composition of Antarctic Phaeocystis sp. (Prymnesiophyceae). II. Pigment composition. J. Phycol. 1998, 34, 496–503.

197. Kennish, M.J. Ecology of Estuaries: Anthropogenic Effects; CRC Press: Boca Raton, FL, USA, 1992.

198. Campanella, L.; Cubadda, F.; Sammartino, M.; Saoncella, A. An algal biosensor for the monitoring of water toxicity in estuarine environments. Water Res. 2001, 35, 69–76.

199. Crist, R.H.; Martin, J.R.; Guptill, P.W.; Eslinger, J.M.; Crist, D.L.R. Interaction of metals and protons with algae. 2. Ion exchange in adsorption and metal displacement by protons. Environ. Sci. Technol. 1990, 24, 337–342.

200. Rai, L.; Mallick, N. Heavy metal toxicity to algae under synthetic microcosm. Ecotoxicology 1993, 2, 231–242.

201. Stauber, J.; Florence, T. Mechanism of toxicity of ionic copper and copper complexes to algae. Mar. Biol. 1987, 94, 511–519.

202. Peterson, H.G.; Healey, F.P.; Wagemann, R. Metal toxicity to algae: A highly pH dependent phenomenon. Can. J. Fish. Aquat. Sci. 1984, 41, 974–979.

203. Wong, P.; Burnison, G.; Chau, Y. Cadmium toxicity to freshwater algae. Bull. Environ. Contam. Toxicol. 1979, 23, 487–490.

204. Wong, P.; Chau, Y. Zinc toxicity to freshwater algae. Toxic. Assess. 1990, 5, 167–177.

205. Sterner, R.W.; Grover, J.P. Algal growth in warm temperate reservoirs: Kinetic examination of nitrogen, temperature, light, and other nutrients. Water Res. 1998, 32, 3539–3548.

206. Smith, A.; Morris, I. Pathways of carbon assimilation in phytoplankton from the Antarctic Ocean. Limnol. Oceanogr. 1980, 25, 865–872.

207. Kudo, I.; Miyamoto, M.; Noiri, Y.; Maita, Y. Combined effects of temperature and iron on the growth and physiology of the marine diatom Phaeodactylum tricornutum (Bacillariophyceae). J. Phycol. 2000, 36, 1096–1102.

208. Cloern, J.E.; Grenz, C.; Vidergar-Lucas, L. An empirical model of the phytoplankton chrlorphyll: Carbon ratio—the conversion factor between productivity and growth rate. Limnol. Oceanogr. 1995, 40, 1313–1321.

209. Morgan, K.C.; Kalff, J. Effect of light and temperature interactions on growth of Cryptomonas erosa (Cryptophyceae). J. Phycol. 1979, 15, 127–134.

210. Sorokin, C.; Krauss, R.W. Effects of temperature & illuminance on chlorella growth uncoupled from cell division. Plant Physiol. 1962, 37, 37–42.

211. Parker, M.S.; Armbrust, E. Synergistic effects of light, temperature, and nitrogen source on transcription of genes for carbon and nitrogen metabolism in the centric diatom Thalassiosira pseudonana (Bacillariophyceae). J. Phycol. 2005, 41, 1142–1153.

212. Cunningham, F.X., Jr.; Dennenberg, R.J.; Jursinic, P.A.; Gantt, E. Growth under red light enhances photosystem II relative to photosystem I and phycobilisomes in the red alga Porphyridium cruentum. Plant Physiol. 1990, 93, 888–895.

213. Cordell, D.; Drangert, J.O.; White, S. The story of phosphorus: Global food security and food for thought. Glob. Environ. Chang. 2009, 19, 292–305.

214. Yang, P.Y.; Duerr, E.O. Bio-process of anaerobically digested pig manure for production of Spirulina sp. Am. Soc. Agric. Eng. 1987. fiche no. 87-6056.

215. Wilkie, A.C.; Mulbry, W.W. Recovery of dairy manure nutrients by benthic freshwater algae. Bioresour. Technol. 2002, 84, 81–91.

216. Dodd, J.C. Algae production and harvesting from animal wastewaters. Agric. Wastes 1979, 1, 23–37.

217. An, J.Y.; Sim, S.J.; Lee, J.S.; Kim, B.W. Hydrocarbon production from secondarily treated piggery wastewater by the green alga Botryococcus braunii. J. Appl. Phycol. 2003, 15, 185–191.

218. Aslan, S.; Kapdan, I.K. Batch kinetics of nitrogen and phosphorus removal from synthetic wastewater by algae. Ecol. Eng. 2006, 28, 64–70.

219. Bich, N.N.; Yaziz, M.I.; Bakti, N.A.K. Combination of Chlorella vulgaris and Eichhornia crassipes for wastewater nitrogen removal. Water Res. 1999, 33, 2357–2362.

220. González, L.E.; Cañizares, R.O.; Baena, S. Efficiency of ammonia and phosphorus removal from a colombian agroindustrial wastewater by the microalgae Chlorella vulgaris and Scenedesmus dimorphus. Bioresour. Technol. 1997, 60, 259–262.

221. Martınez, M.E.; Sánchez, S.; Jiménez, J.M.; El Yousfi, F.; Muñoz, L. Nitrogen and phosphorus removal from urban wastewater by the microalga Scenedesmus obliquus. Bioresour. Technol. 2000, 73, 263–272.

222. Hoffmann, J.P. Wastewater treatment with suspended and nonsuspended algae. J. Phycol. 1998, 34, 757–763.

223. Merchant, S.S.; Prochnik, S.E.; Vallon, O.; Harris, E.H.; Karpowicz, S.J.; Witman, G.B.; Terry, A.; Salamov, A.; Fritz-Laylin, L.K.; Maréchal-Drouard, L. The Chlamydomonas genome reveals the evolution of key animal and plant functions. Science 2007, 318, 245–250.

224. Blanc, G.; Duncan, G.; Agarkova, I.; Borodovsky, M.; Gurnon, J.; Kuo, A.; Lindquist, E.; Lucas, S.; Pangilinan, J.; Polle, J. The Chlorella variabilis NC64A genome reveals adaptation to photosymbiosis, coevolution with viruses, and cryptic sex. Plant Cell Online 2010, 22, 2943–2955.

225. Smith, D.; Lee, R.; Cushman, J.; Magnuson, J.; Tran, D.; Polle, J. The Dunaliella salina organelle genomes: Large sequences, inflated with intronic and intergenic DNA. BMC Plant Biol. 2010, 10, 83–96.

226. Chang, R.L.; Ghamsari, L.; Manichaikul, A.; Hom, E.F.; Balaji, S.; Fu, W.; Shen, Y.; Hao, T.; Palsson, B.Ø.; Salehi-Ashtiani, K.; et al. Metabolic network reconstruction of Chlamydomonas offers insight into light-driven algal metabolism. Mol. Syst. Biol. 2011, 7.

10
CHAPTER
BIOFUELS AND ECOSYSTEM CARBON BALANCE UNDER GLOBAL CHANGE

Christina Schädel [1] and Yiqi Luo[1]

[1] Department of Botany and Microbiology, University of Oklahoma, USA

1. INTRODUCTION

Terrestrial ecosystems are estimated to contain 3000 Pg of organic carbon (C) of which more than two thirds are stored in soils (Jobbágy and Jackson 2000). Total soil organic carbon (SOC) consists of different C pools with intrinsic turnover rates ranging from less than a year to thousands of years (Amundson 2001). The processes that drive the cycling of soil C are C inputs from net primary productivity (NPP=GPP-ecosystem respiration) and outputs through C decomposition (Fig. 1). New organic matter is the product of NPP which is transferred to soils in the form of litter and presents the largest C input to soils. Carbon output to the atmosphere is mainly driven by soil respiration (consisting of autotrophic and heterotrophic respiration) which is the second largest driver of C in the global C cycle (Fig. 1). Annually, soils release 98 ± 12 Pg C to the atmosphere which has increased yearly by 0.1 Pg C between 1989-2008 (Bond-Lamberty and Thomson 2010) and which yearly exceeds the current rate of fossil fuel combustion by a factor of 10. These large numbers show that even slight changes in the soil C and soil C cycling are highly relevant to the global C cycle mainly because of their potential to sequester or release CO_2 (Trumbore 1997). C sequestration denotes the transfer of C from atmospheric CO_2 into long-lived pools (e.g. woody biomass, recalcitrant soil C pools) without reemitting it immediately.

Although the soil is a dynamic system C input and output need to be balanced in order to keep the SOC pool at equilibrium (Fig. 2a). If C input is smaller than C output a depletion of the SOC pool occurs which can result in large releases of CO_2 to the atmosphere (Fig. 2b). On the other hand, if C input to the soil exceeds C output additional SOC can be sequestered in soils. SOC is not only an important C sink within the terrestrial C budget it also strongly influences soil fertility and soil quality which in return is needed for plant growth (Lal 2004; Cruse et al. 2010).

Global change denotes all human-caused changes to the atmosphere, hydrosphere, pedosphere and biosphere (Körner 2003). The increasing CO_2 concentration in the atmosphere is one of the most drastic global change components that directly affect plants and the ecosystems they live in (IPCC 2007). Secondary effects of higher CO_2 concentrations are climate warming causing tertiary effects such as extended growing seasons, shifts in species composition and alterations in precipitation patterns. Elevated CO_2 directly affects plants through photosynthesis and as photosynthesis at the current level of CO_2 concentration is not yet CO_2-saturated there is leeway for more carbon fixation and with it the possibility of more C storage in terrestrial ecosystems. On the other hand climate warming affects almost all aspects of carbon cycling and enhanced C fluxes potentially feed back to the atmosphere causing the so called positive feedback to climate change (Luo 2007).

Identifying possible strategies for mitigating climate change by reducing increases in atmospheric CO_2 has put a strong focus on growing biofuel feedstock for alternative energy. Biofuels are generated through the combustion of biomass, usually grain or cellulosic-based feedstock. Biofuel feedstock production can help offset C emissions from fossil fuels but continuous biomass harvesting involves the removal of large quantities of C inducing a disequilibrium in the ecosystem's C balance (Fig. 2b; Luo and Weng 2011). There is thus an urgent need to investigate the impacts of biofuel feedstock harvesting on an ecosystem's C balance and its feedback to climate change (Luo et al. 2009)

This chapter focuses on key issues related to biofuel feedstock harvesting and ecosystem C balance under global change.

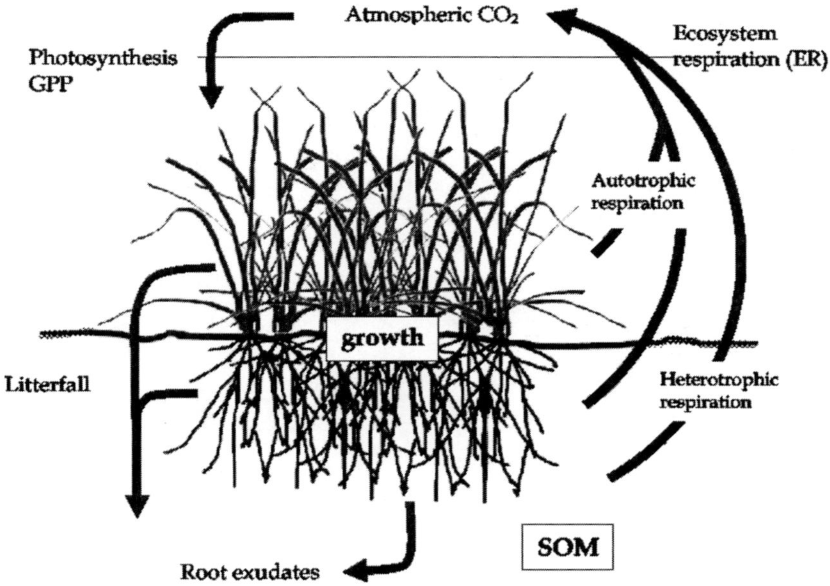

Figure 1. Schematic diagram of terrestrial ecosystem C processes. GPP, gross primary production, SOM, soil organic matter

2. TERRESTRIAL ECOSYSTEM C CYCLE

Carbon enters a terrestrial ecosystem through photosynthetic uptake (GPP=gross primary production) and is either incorporated into biomass, respired through ecosystem respiration (ER=autotrophic + heterotrophic respiration) or secreted into the soil via litter and root exudates (Fig. 1; Chapin et al. 2009). The main terrestrial ecosystem C cycling processes are photosynthesis and C decomposition and hence are the main components to calculate net ecosystem productivity (NEP=GPP-ER). Once C has entered the plant system it is allocated to above- and belowground tissues with the partitioning being strongly dependent on nutrient availability. The partitioning of C allocation within the plant influences how much C goes into aboveground biomass (leaves, stems) and how much enters the soil (coarse and fine roots). Belowground primary productivity can account for up to 33% of the annual net primary productivity (Jackson et al. 1997; Norby et al. 2004) which assigns roots an important role in the net ecosystem C balance. Dead plant material from above- and belowground litter is decomposed by microorganisms releasing CO_2 to the atmosphere and providing energy for microbial biomass growth. Microbial biomass together with organic residuals of dead plant material and dead microbes form the soil organic matter (SOM, Luo and Zhou 2006) of which carbon represents 50% by weight. C decomposition from short- and long-lived C pools is one of the most important factors in regulating terrestrial ecosystem C cycles.

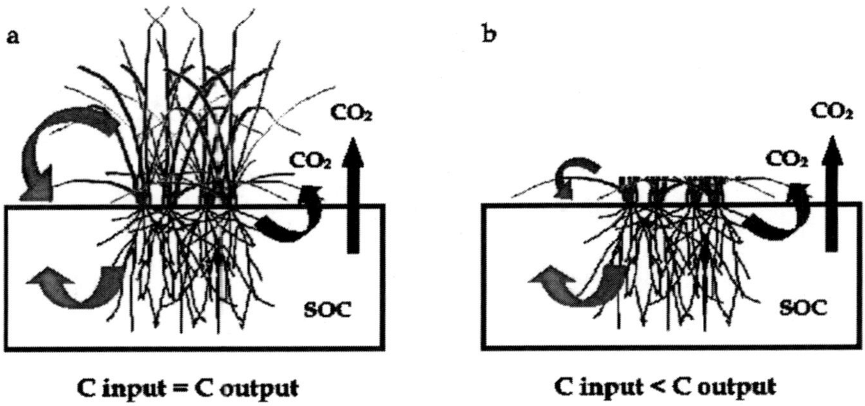

Figure 2. Soil C dynamics for a) a natural grassland and b) a harvested field for biofuel feedstock. Red arrows represent C input to the soil and black arrows C output

3. GLOBAL CHANGE AND C PROCESSES

Global change is one of the most important issues that this century is going to face. Global atmospheric CO_2 concentrations gradually increased from pre-industrial concentrations of 280 ppm to 390 ppm in 2011 (Mauna Loa, Hawaii; Tans 2011) and continue to increase, the main causes being deforestation, land use change and the burning of fossil-fuels (IPCC 2007). As a consequence to the increasing CO_2 and other green house gas concentrations in the atmosphere, global temperatures have increased by 0.74°C over the last century and predictions for the end of this century suggest a further increase in temperatures by another 1.1°C - 6.4°C (IPCC 2007). Temperature affects almost all chemical and biological processes (Shaver et al. 2000) and increasing global temperatures will likely enhance C fluxes between pools. As primary productivity and C decomposition are directly affected by changes in temperature ecosystem C uptake will be highly influenced by global warming. One of the major uncertainties of global warming is a possible positive feedback between the global carbon cycle and climate change but interactions among processes and fluxes are most uncertain making predictions difficult (Shaver et al. 2000; Luo 2007). Global change also includes altered precipitation regimes as a direct result to climate warming with a higher frequency in extreme events (rainfall extreme, drought) being very likely (IPCC 2007). Furthermore, human activities have increased the rate and magnitude of N deposition (Galloway et al. 2004) and since N availability strongly influence C processes future ecosystem C sequestration largely depends on N availability (Luo et al. 2004; Luo et al. 2006; Field et al. 2007). Global change includes various factors and their multiple interactions have smaller or larger impacts on ecosystem C processes and understanding those interactive effects on ecosystem processes becomes crucial in predicting possible carbon-climate feedbacks (Luo et al. 2008).

3.1. Impacts of global change on C processes

C influx through photosynthesis is stimulated by increased CO_2 concentrations (Curtis and Wang 1998; Norby et al. 1999; Ainsworth and Long 2005) often leading to increasing biomass growth but results are highly variable between species and plant age (Norby et al. 1999; Körner et al. 2005). Higher CO_2 concentrations could also induce higher C sequestration in soils (Jastrow et al. 2005; Lichter et al. 2005) but results strongly varied with CO_2 experimental facility, ecosystem type and N treatment (Luo et al. 2006). Highest C sequestration rates in soils were found in studies with N fertilization (Hungate et al. 2009) and most likely reflect effects of elevated CO_2 and N on plant growth as well as interactions between CO_2 and N on soil C decomposition (Reich et al. 2006). Increased soil C efflux under elevated CO_2 was reported from most CO_2-enrichment studies which was attributed to additional substrate from greater plant growth under elevated CO_2 available to soil microorganisms (Zak et al. 2000).

Increasing temperatures affect most biochemical processes such as C influx and C decomposition. Results are highly variable and range from no change, increases and decreases in photosynthesis, plant growth, primary production and soil respiration (Luo 2007). Generally, C decomposition reacts more to higher temperatures (Melillo et al. 2002; Zhou et al. 2007) than does C influx although the response of increasing soil C decomposition to warming attenuates over time (Rustad et al. 2001). The overall response of C decomposition to climate warming could be a net release of CO_2 from soils which might result in a positive feedback between C cycles and climate warming within this century (Friedlingstein et al. 2006). Furthermore, climate warming affects regulatory mechanisms of ecosystem C processes such as increased length of growing season, changes in water dynamics, species composition and nutrient availability (Luo 2007).

Precipitation patterns are changing in its frequency and intensity as a consequence of climate warming. Increased precipitation mostly increases C uptake via photosynthesis as well as increases soil respiration (Zhou et al. 2009). Drier conditions as they occur during heat waves reduce gross primary productivity and result in large net releases of CO_2 (Ciais et al. 2005; Arnone et al. 2008). This net release in CO_2 can offset C sequestration from previous years and turn ecosystems into C sources which will then contribute to a positive carbon-climate feedback. Besides the direct effects of increasing or decreasing water availability there are indirect effects of water availability through other climate change factors. Elevated CO_2 usually leads to less water loss through stomata (Medlyn et al. 2001; Leuzinger and Körner 2007) whereas a warmer climate increases evaporation leading to more negative water balances in the soil. As primary production in many ecosystems is largely limited by water availability any changes in precipitation will have substantial impacts on ecosystem C dynamics (Heimann and Reichstein 2008).

Nitrogen is one of the most limiting nutrients for plant growth (Vitousek and Howarth 1991) and N fertilization is widely used to improve plant productivity. N addition can increase plant growth by 30-50% although there are large differences between plant functional types and biomes (LeBauer and Treseder 2008; Xia and Wan 2008). Interactions between C and N cycles are important as N has the potential to sequester C when plant growth is not restricted by the lack of available N. Aboveground biomass is more stimulated by additional N availability than belowground biomass reducing C input to the soil (Lu et al. 2011). Litter quality usually increases with N addition which enhances decomposition of this high-quality litter (Knorr et al. 2005) causing less accumulation of C in the soil. Results on soil C storage under N addition are controversial and likely vary between natural ecosystems and agricultural ecosystems (Pregitzer et al. 2008; Liu and Greaver 2010; Lu et al. 2011). As climate change is a multi-factor process we need to investigate interactions between those climate change factors and expect dynamics that we cannot investigate by only looking at one factor.

4. IMPACTS OF BIOFUEL FEEDSTOCK HARVESTING ON C CYCLES

4.1. Biomass removal and SOC

Crop and plant residues (usually deriving from corn, barley, oat, sorghum, soybean, sunflower and wheat) are considered to be free products for biofuel production as they are left behind after harvest and seem not be of any use (Lal 2004). Plant residues are high-cellulosic feedstock (high concentration of cellulose and hemicelluloses) which are suitable for ethanol production but are not in direct competition to food compared to starch-rich biomass from grain. Complete biomass removal might impact soil quality, SOC content and soil water content more than a partial removal. The amount of corn stover (referring to all aboveground plant material such as stalk, leaves, cobs and husk) residue that can be removed without endangering SOC contents of soils was estimated to be 25% of total removal (Blanco-Canqui and Lal 2007). If more than 25% of corn stover was removed soils showed decreasing SOC contents.

Most biofuel feedstock derives from monocultures such as corn, soybeans, switchgrass or sugarcane and is grown on fertile soils. However, studies have shown that more bioenergy can be produced from switch-grass mixtures or high-diversity grasslands than from monocultures (Tilman et al. 2006; Wang et al. 2010). Studies have shown that even though aboveground biomass of high-diversity grasslands was removed annually soils still sequestered carbon. The full life-cycle of these high-diversity grasslands results in a net sequestration of atmospheric CO_2, which makes these high-diversity grasslands so called carbon negative biofuels.

The partitioning of C allocation within the plant influences how much C goes into harvestable biomass and how much enters the soil. Depending on the vegetation type of the biofuel feedstock C allocation strongly varies and influences the amount of C stored in soils. Root architecture and their vertical distribution contribute to soil C maintenance with less C stored in deeper soil layers. Deep soil layers though store the highest amounts of highly stable organic C compounds and are less affected by freshly added litter input. It was suggested that in the absence of fresh organic carbon input the stability of organic compounds in deep soil layers is maintained as the available energy required to breakdown recalcitrant SOC is not provided (Fontaine et al. 2007). Therefore biomass removal for biofuel production could protect deep soil recalcitrant C pools from decomposition.

Root systems play as much a role in the build-up and maintenance of SOC as does the amount of residue removed. Especially graminoids (to which most biofuel crops belong to) have larger ratios of belowground biomass to aboveground biomass and therefore allocate more C through belowground biomass to the soil than through aboveground biomass which gives them a higher C sequestration potential (De Deyn et al. 2008). It was even suggested that belowground biomass is more important for C sequestration than aboveground biomass as studies showed that changes in SOC pools positively correlated with the quantity of belowground biomass input but not with input of aboveground biomass (Russell et al. 2009; Lu et al. 2011). Balesdent and Balabane (1996) measured root-derived C in maize cultivated soils and found that although the shoot to root ratio was only 0.5 root-derived C was 1.5 times higher than aboveground-derived C (from stalks and leaves). Furthermore, root litter of grasses is of lower quality and therefore less easily decomposable compared to aboveground litter due to lower N but higher lignin concentration (Vivanco and Austin 2006). This higher recalcitrance of plant litter slows down the litter decay process and increases the amount of C stored in the soil (Sartori et al. 2006; Johnson et al. 2007).

4.2. Biofuel feedstock harvest and global change

The sustainability of biofuel feedstock harvest under global change needs to be evaluated in order to quantify changes in the net ecosystem C balance as well as assess a possible positive feedback to climate change. Biofuel feedstock harvest and the coherent changes in the C balance can be evaluated from experimental studies that use clipping or biomass harvesting to remove aboveground biomass (Luo et al. 2009). One study that combined the effects of climate warming and biomass feedstock harvesting on ecosystem C dynamics was conducted in the Southern Great Plains, USA, which is considered to be a major region for biofuel feedstock production (Luo et al. 2009). Temperatures were increased by 2°C and biomass was clipped annually. On average, data of nine years showed increased net primary productivity (NPP) under warming and even higher values in the combination treatment of warming and clipping. Although warming increased soil respiration rates clipping showed a decreasing trend in soil respiration. Yearly biomass removal reduced the C input to soils which was clearly demonstrated by higher losses of soil C in the clipped plots compared to the unclipped plots (Fig. 3). In both clipped treatments losses in soil C after nine years were more than twice as high as they were for the unclipped plots. Additionally, warming enhanced soil C loss resulting in the highest loss of soil C under clipping and warming treatment (Fig. 3). These results clearly show that biofuel feedstock harvest in combination with warmer temperatures results in the highest loss in soil C.

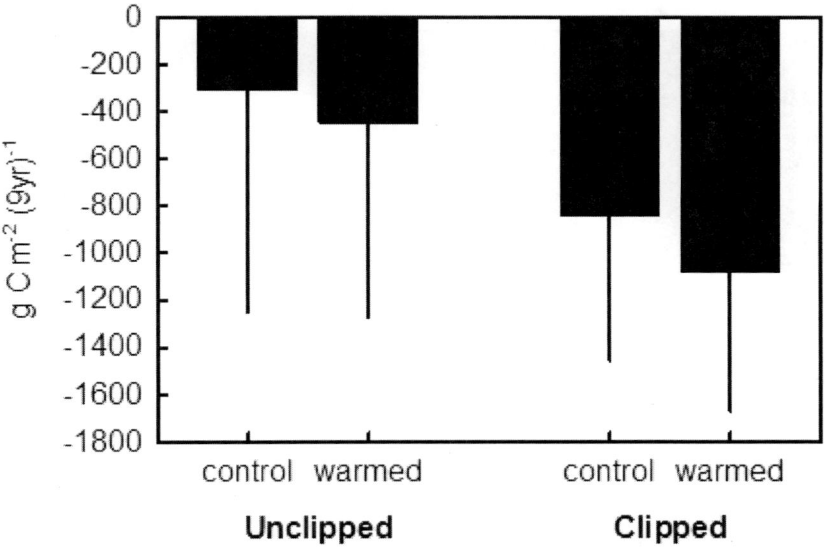

Figure 3. Change in soil C content between 1999-2008. Values are means of 5 plots ± 1 se

4.3. Clipping-induced erosion under global change

Changes in land use through alteration of land coverage and disturbance of soil structure result in changes in soil moisture which can induce higher soil erosion rates (Lal 2004). Generally, plant coverage protects the soil from soil erosion by intercepting rainfall and runoff. Plant cover, plant height, rooting characteristics and other plant related parameters are important factors in reducing soil erosion rates (Wilhelm et al. 2007; Johnson et al. 2010). If aboveground biomass is removed for biofuel feedstock harvest more bare ground will increase temperatures as well as surface runoff and thus accelerate soil erosion (Schlesinger et al. 1990; Zuazo and Pleguezuelo 2008). Cover and type of vegetation can therefore affect soil erosion and potentially lead to a net source of C by soil erosion induced loss of SOC.

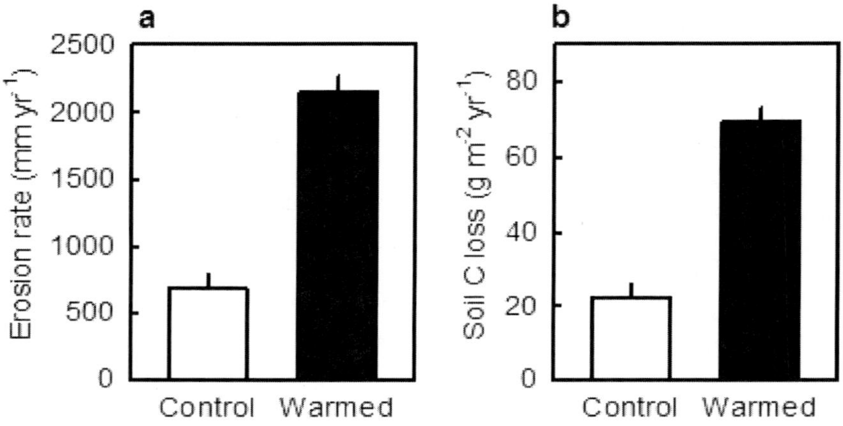

Figure 4. a) Yearly erosion rate in the clipped subplots, b) yearly soil C loss in the clipped subplots. Values are means of 16 measurements per treatment ± SD. Redrawn with permission from *Global Change Biology Bioenergy*, Xue et al. 2011

It is well known that biomass removal on a continuous basis results in increased soil erosion but it is not well known how a warmer climate might amplify C loss from soils through erosion. The only study, we are aware of, that combines the effects of biomass removal and climate warming on soil erosion rates was conducted in a tallgrass prairie in the Southern Great Plains, USA (Xue et al. 2011). In a multiyear experiment (since 1999) grassland was warmed (+2°C) on a whole ecosystem-level and half the plots were clipped in order to mimic biofuel feedstock harvest. One side effect of warming was a reduction in soil moisture which was even greater in the clipped plots. Clipping-induced relative soil erosion rate was threefold increased under the warming scenario (Fig. 4a). These high erosion rates resulted in high losses of SOC (Fig. 4b). The stronger response to the warming treatment in the clipped plots was ascribed to lower soil moisture in the clipped plots as evaporation from the soil surface was increased when biomass was removed. Some of the consequences of higher erosion rates are reduced soil fertility, degraded soil structure and reduced SOC, all being enhanced by biomass removal. The soil that is most affected by erosional processes is the SOC-rich upper soil level making erosion a net source of C to the atmosphere (e.g. Lal 2003).

5. INTERACTIVE EFFECTS OF BIOFUEL FEEDSTOCK HARVESTING AND GLOBAL CHANGE

5.1. Biofuel feedstock harvesting and NECB

Soils and their C stocks will be affected by land use change and by manipulations in the substrate supply but more importantly changes in the soil C budget will potentially affect the net ecosystem C balance (Fargione et al. 2008; Sanderson 2008; Luo et al. 2009) and consequentially contribute to the overall terrestrial C-cycle feedback.

Ecosystems can function as C sources or C sinks and their role in the global C cycle becomes even more important with global change as ecosystems either release or absorb atmospheric CO_2 and with it enhance or mitigate climate warming (Chapin et al. 2006). Net ecosystem production (NEP) is a measure of gross primary productivity (GPP) minus ecosystem respiration and mostly coincides with the net ecosystem C balance (NECB) unless C in other forms than CO_2 or dissolved organic C moves in or out of the system (Chapin et al. 2006; Lovett et al. 2006). Therefore, NECB is the net estimate of C accumulation (positive NECB) or C loss (negative NECB) in any system. If an ecosystem's net C balance is positive the ecosystem functions as a C sink by sequestering C. In contrast, a negative NECB implies C release to the atmosphere and any ecosystem showing a negative balance functions as a C source. NECB can be applied on short-term or long-term scales and to any spatial scale which makes it a very useful parameter for cross-scale comparisons (Chapin et al. 2006). To fully estimate the impact of biofuel feedstock removal on ecosystems under global change the net ecosystem C balance needs to be calculated to estimate a feedback of biomass removal to climate change. So far there are not many studies that measure the impacts of biofuel feedstock harvest on the net ecosystem C balance under global change. Nevertheless this is important as biofuels are supposed to help mitigate climate change by reducing CO_2 release from fossil fuels. But if biofuel feedstock harvest has large negative impacts on the net ecosystem C balance this mitigation strategy might not help reduce CO_2 release to the atmosphere.

5.2. NECB under elevated CO_2

Elevated atmospheric CO_2 generally increases above- as well as belowground biomass and also enhances soil C storage although the extent to which C is stored in soils is largely dependent on N availability (Luo et al. 2006). Belowground biomass often shows a higher response to elevated

CO_2 therefore increasing C input to soils (Luo et al. 2006). C accumulation in plant and soil pools reflects increased C input into ecosystems that usually decreases litter quality and with it decomposability. Decreasing decomposability also derives from increased mycorrhizal growth under elevated CO_2 that enhances physical protection through formation of intra-aggregate or organomineral complexes to protect organic matter from microbial decomposition (Rillig 2004). Large fractions of the C accrued in soils under elevated CO_2 derive from increased belowground biomass growth which is not affected by biomass removal. Nevertheless there are some factors that need to be considered when making predictions about net ecosystem C balances for biofuel feedstock harvest under elevated CO_2. It is not yet clear whether there will be a down-regulation of CO_2 stimulation of photosynthesis and with it in plant growth and other C processes under persistent CO_2 stimulation (Long et al. 2004;). Photosynthetic acclimation was alleviated in grassland when plants were harvested but only under high N availability (Ainsworth et al. 2003). Low N conditions resulted in some acclimation of photosynthetic capacity. It seems that all responses of C processes under elevated CO_2 are strongly dependent on N availability. However, when only considering the global change factor elevated CO_2, biofuel feedstock harvest might still allow for C sequestration in soils resulting in a positive net ecosystem C balance.

5.3. NECB under climate warming

Unlike elevated CO_2 that primarily influences C uptake through photosynthesis warming affects almost all chemical and biological processes. Furthermore, warming involves some secondary effects on ecosystems such as extended growing seasons, change in species composition and drier conditions. Hence, it is not surprising that ecosystem warming experiments have produced inconsistent results regarding plant growth, soil respiration and net ecosystem production. Nevertheless the most important biomass fraction for C sequestration under biofuel feedstock harvest is the belowground biomass which was positively stimulated under warming and harvesting scenarios (Luo et al. 2009). This positive interaction was ascribed to over-compensatory mechanisms of plant physiological processes to biomass removal (Owensby et al. 2006). As belowground biomass growth is enhanced under warmer conditions the C loss through biomass removal might be less important for the net ecosystem C balance than the gain in C through increased belowground biomass. On the other hand continuous biomass removal increases soil erosion rates (Xue et al. 2011) which is accompanied by high losses of soil C. Even higher erosion rates occur when biomass removal takes place under warmer conditions as the soil dries out more easily leaving unstable soil structures favoring soil erosion. Therefore, biomass harvesting of natural grassland (Luo et al. 2009) in combination with warming resulted in a more negative net ecosystem C balance than for the warming treatment alone (Fig. 5). The more negative C balance is mainly due to high soil C losses (Fig. 4) as C input to soils was smaller than the C lost through CO_2 release and soil erosion. Thus, over-compensatory belowground biomass growth was not enough to offset soil C loss under warming and clipping. This long-term experiment shows that growing biofuel feedstock for harvesting under climate warming puts an additionally strain on the ecosystem C balance and does not help to sequester more C in order to reduce CO_2 release to the atmosphere.

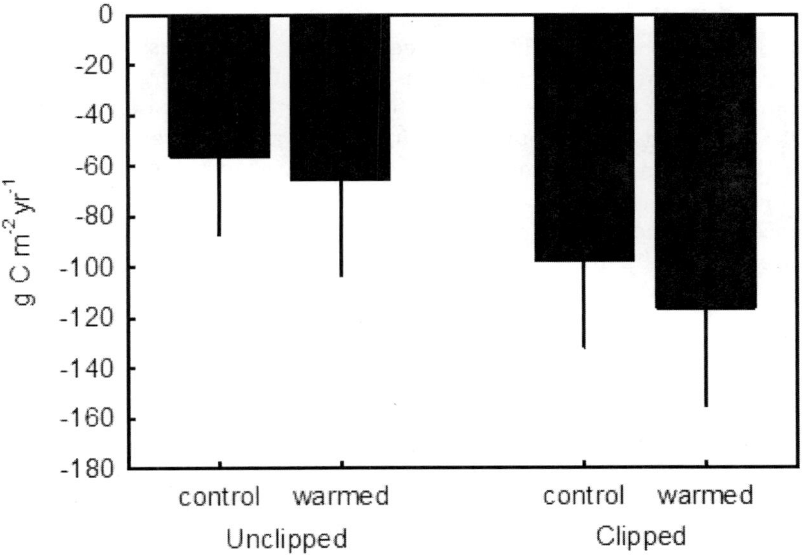

Figure 5. Net ecosystem C balance calculated per year for the period of 2000-2008. Values are means of 6 plots ± 1 se

5.4. NECB and change in precipitation

Changes in precipitation as a consequence of global change include more frequent extreme precipitation and drought events which likely have large effects on ecosystem processes (Weltzin et al. 2003). Precipitation is an important factor in shaping ecosystem C dynamics as aboveground biomass and soil respiration linearly increase with mean annual precipitation but belowground biomass and soil C content remain rather constant (Zhou et al. 2009). As was shown for the Southern Great Plains in the USA no change in belowground C allocation is more important to the net ecosystem C balance than higher aboveground plant growth since this higher aboveground litter input was compensated by higher litter decomposition. A more positive net ecosystem C balance therefore seems plausible under wetter conditions. On the other hand warming induced drought suppresses net primary productivity and turns ecosystems into net sources of carbon dioxide (Ciais et al. 2005; Arnone et al. 2008). If additionally biomass is removed the net ecosystem C balance could become even more negative contributing more to a positive carbon-climate feedback.

5.5. NECB and N addition

N addition strongly influences ecosystem C processes through photosynthesis and biomass production and therefore has large impacts on the net ecosystem C balance. Generally N addition increases C input to soil through increased aboveground litter input (Liu and Greaver 2010). With higher N availability plants invest less C into belowground biomass as roots can more easily acquire N. Furthermore, higher N availability strongly influences the shoot to root ratio and root litter flux to soil decreases (Liu and Greaver 2010). If additionally C from aboveground biomass is not returned to soil due to biofuel feedstock harvest total C input to the soil will decrease and a negative net ecosystem C balance is very likely.

6. CONCLUSION

Growing biofuels for alternative energy can help mitigate increasing atmospheric CO_2 concentration; however continuous biofuel feedstock harvest will influence the whole ecosystem C balance possibly resulting in a positive feedback to climate change. Ecosystem C processes are strongly influenced by global change factors and their interactive effects are very complex and not yet well understood. An overall response of biomass feedstock removal on the net ecosystem C balance under global change is therefore still speculative but we know that global change factors that enhance root biomass have a more positive effect on the net ecosystem C balance when biomass is continuously removed than factors that enhance aboveground biomass. Increased CO_2 concentration in the atmosphere has the potential to increase belowground C storage especially when N and other nutrients are not limiting. Climate warming on the other hand seems to reduce soil C storage as C decomposition and C losses through soil erosion under biofuel feedstock harvest are higher. Responses to changes in precipitation are very variable but drier conditions result in a more negative ecosystem C balance if biomass is continuously removed. This effect could be neutralized again under elevated CO_2 as stomatal conductance and evapotranspiration decline thus decreasing the plant water use. N availability is a crucial factor for optimized plant growth and C storage but high N addition can also reduce belowground biomass and thus C input to soils. If additionally all biomass is removed there will be an even smaller C input into soil. One way to alleviate strong impacts of biomass harvest on C-cycling might be to harvest at a later time as harvesting after plant senescence showed to reduce C and N losses although biomass yield might be slightly lower (Heaton et al. 2009; Niu et al. 2010).

In conclusion, this chapter showed that biofuel harvesting has large impacts on the net ecosystem C balance which are likely enhanced under global change. More information on interactive effects of multiple global change factors is still needed to fully estimate the impacts of biofuel feedstock harvest on net ecosystem C balance and any possible feedback to climate change.

REFERENCES

1. E. A. Ainsworth, P. A. Davey, G. J. Hymus, C. P. Osborne, A. Rogers, H. Blum, J. Nösberger, S. P. Long, 2003Is stimulation of leaf photosynthesis by elevated carbon dioxide concentration maintained in the long term? A test with Lolium perenne grown for 10 years at two nitrogen fertilization levels under Free Air CO2 Enrichment (FACE). Plant Cell and Environment, 26 5 705 714

2. E. A. Ainsworth, S. P. Long, 2005What have we learned from 15 years of free-air CO2 enrichment (FACE)? A meta-analytic review of the responses of photosynthesis, canopy properties and plant production to rising CO2. New Phytologist, 165 2 351 371

3. R. Amundson, 2001The carbon budget in soils. Annual Review of Earth and Planetary Sciences, 29 535 562

4. J. A. Arnone, et al. 2008Prolonged suppression of ecosystem carbon dioxide uptake after an anomalously warm year. Nature, 455 7211 383 386

5. J. Balesdent, M. Balabane, 1996Major contribution of roots to soil carbon storage inferred from maize cultivated soils. Soil Biology & Biochemistry, 28 9 1261 1263

6. H. Blanco-Canqui, R. Lal, 2007Soil and crop response to harvesting corn residues for biofuel production. Geoderma, 141, 3-4, 355-362

7. B. Bond-Lamberty, A. Thomson, 2010Temperature-associated increases in the global soil respiration record. Nature, 464 7288 579 582

8. F. S. Chapin, J. Mc Farland, A. D. Mc Guire, E. S. Euskirchen, R. W. Ruess, K. Kielland, 2009The changing global carbon cycle: linking plant-soil carbon dynamics to global consequences. Journal of Ecology, 97 5 840 850

9. F. S. Chapin, et al. 2006Reconciling carbon-cycle concepts, terminology, and methods. Ecosystems, 9 7 1041 1050

10. Ciais, P. et al. (2005). Europe-wide reduction in primary productivity caused by the heat and drought in 2003. Nature, 437, 7058, 529-5337058 529 533

11. R. M. Cruse, M. J. Cruse, D. C. Reicosky, 2010Soil quality impacts of residue removal for biofuel feedstock. In: Soil quality and biofuel production. Lal, R. and Stewart, B. A., (Ed.) CRC Press, Taylor & Francis Group, 978-1-43980-073-7Boca Raton, FL

12. P. S. Curtis, X. Z. Wang, 1998A meta-analysis of elevated CO2 effects on woody plant mass, form, and physiology. Oecologia, 113 3 299 313

13. G. B. De Deyn, J. H. C. Cornelissen, R. D. Bardgett, 2008Plant functional traits and soil carbon sequestration in contrasting biomes. Ecology Letters, 11 5 516 531

14. J. Fargione, J. Hill, D. Tilman, S. Polasky, P. Hawthorne, 2008Land clearing and the biofuel carbon debt. Science, 319 5867 1235 1238

15. C. B. Field, D. B. Lobell, H. A. Peters, N. R. Chiariello, 2007Feedbacks of terrestrial ecosystems to climate change. Annual Review of Environment and Resources, 32 1 29

16. S. Fontaine, S. Barot, P. Barré, N. Bdioui, B. Mary, C. Rumpel, 2007Stability of organic carbon in deep soil layers controlled by fresh carbon supply. Nature, 450 7167 277U210

17. P. Friedlingstein, et al. 2006Climate-carbon cycle feedback analysis: Results from the (CMIP)-M-4 model intercomparison. Journal of Climate, 19 14 3337 3353

18. J. N. Galloway, et al. 2004Nitrogen cycles: past, present, and future. Biogeochemistry, 70 2 153 226

19. E. A. Heaton, F. G. Dohleman, S. P. Long, 2009Seasonal nitrogen dynamics of Miscanthus x giganteus and Panicum virgatum. Global Change Biology Bioenergy, 1 4 297 307

20. M. Heimann, M. Reichstein, 2008Terrestrial ecosystem carbon dynamics and climate feedbacks. Nature, 451 7176 289 292

21. Hungate, B. A.; van Groenigen, K. J.; Six, J.; Jastrow, J. D.; Lue, Y. Q.; de Graaff, M. A.; van Kessel, C. & Osenberg, C. W. (2009). Assessing the effect of elevated carbon dioxide on soil carbon: a comparison of four meta-analyses. Global Change Biology, 15, 8, 2020-20348 2020 2034

22. IPCC (2007). Climate Change 2007: The physical science basis. Contribution of working group I to the fourth assessment report of the Intergovernmental Panel on Climate Change, Cambridge University Press, Cambridge, United Kingdom and New York, NY, USA

23. R. B. Jackson, H. A. Mooney, E. D. Schulze, 1997A global budget for fine root biomass, surface area, and nutrient contents. Proceedings of the National Academy of Sciences of the United States of America, 94 14 7362 7366

24. Jastrow, J. D.; Miller, R. M.; Matamala, R.; Norby, R. J.; Boutton, T. W.; Rice, C. W. & Owensby, C. E. (2005). Elevated atmospheric carbon dioxide increases soil carbon. Global Change Biology, 11, 12, 2057-206412 2057 2064

25. E. G. Jobbágy, R. B. Jackson, 2000The vertical distribution of soil organic carbon and its relation to climate and vegetation. Ecological Applications, 10 2 423 436

26. J. M. E. Johnson, S. K. Papiernik, M. M. Mikha, K. A. Spokas, M. D. Tomer, S. L. Weyers, 2010Soil processes and residue harvest management. In: Soil Quality and Biofuel Production. Lal, R. and Stewart, B. A., (Ed.): 1 44CRC Press, Taylor & Francis Group, 978-1-43980-073-7Boca Raton, FL

27. J. M. F. Johnson, N. W. Barbour, S. L. Weyers, 2007Chemical composition of crop biomass impacts its decomposition. Soil Science Society of America Journal, 71 1 155 162

28. M. Knorr, S. D. Frey, P. S. Curtis, 2005Nitrogen additions and litter decomposition: A meta-analysis. Ecology, 86 12 3252 3257

29. Körner, C. (2003). Ecological impacts of atmospheric CO2 enrichment on terrestrial ecosystems. Philosophical Transactions of the Royal Society of London Series a-Mathematical Physical and Engineering Sciences, 361, 1810, 2023-20411810 2023 2041

30. C. Körner, et al. 2005Carbon flux and growth in mature deciduous forest trees exposed to elevated CO2. Science, 309 5739 1360 1362

31. R. Lal, 2003Soil erosion and the global carbon budget. Environment International, 29 4 437 450

32. R. Lal, 2004Soil carbon sequestration impacts on global climate change and food security. Science, 304 5677 1623 1627

33. D. S. Le Bauer, K. K. Treseder, 2008Nitrogen limitation of net primary productivity in terrestrial ecosystems is globally distributed. Ecology, 89 2 371 379

34. Leuzinger, S. & Körner, C. (2007). Water savings in mature deciduous forest trees under elevated CO2. Global Change Biology, 13, 12, 2498-250812 2498 2508

35. Lichter, J.; Barron, S. H.; Bevacqua, C. E.; Finzi, A. C.; Irving, K. E.; Stemmler, E. A. & Schlesinger, W. H. (2005). Soil carbon sequestration and turnover in a pine forest after six years of atmospheric CO2 enrichment. Ecology, 86, 7, 1835-18477 1835 1847

36. L. L. Liu, T. L. Greaver, 2010A global perspective on belowground carbon dynamics under nitrogen enrichment. Ecology Letters, 13 7 819 828

37. S. P. Long, E. A. Ainsworth, A. Rogers, D. R. Ort, 2004Rising atmospheric carbon dioxide: Plants face the future. Annual Review of Plant Biology, 55 591 628

38. G. Lovett, J. Cole, M. Pace, 2006Is net ecosystem production equal to ecosystem carbon accumulation? Ecosystems, 9 1 152 155

39. M. Lu, X. Zhou, Y. Luo, Y. Yang, C. Fang, J. Chen, B. Li, 2011Minor stimulation of soil carbon storage by nitrogen addition: A meta-

analysis. Agriculture, Ecosystems & Environment, 140, 1-2, 234-244

40. Y. Luo, et al. 2004Progressive nitrogen limitation of ecosystem responses to rising atmospheric carbon dioxide. Bioscience, 54 8 731 739

41. Y. Luo, E. Weng, 2011Dynamic disequilibrium of the terrestrial carbon cycle under global change. Trends in Ecology & Evolution, 26 2 96 104

42. Y. Luo, X. Zhou, 2006 Soil respiration and the environment, Academic Press, 978-0-12088-782-8San Diego, CA, USA

43. Y. Q. Luo, 2007Terrestrial carbon-cycle feedback to climate warming. Annual Review of Ecology Evolution and Systematics, 38 683 712

44. Luo, Y. Q. et al. (2008). Modeled interactive effects of precipitation, temperature, and CO2 on ecosystem carbon and water dynamics in different climatic zones. Global Change Biology, 14, 9, 1986-19999 1986 1999

45. Y. Q. Luo, D. F. Hui, D. Q. Zhang, 2006Elevated CO2 stimulates net accumulations of carbon and nitrogen in land ecosystems: A meta-analysis. Ecology, 87 1 53 63

46. Y. Q. Luo, R. Sherry, X. H. Zhou, S. Q. Wan, 2009Terrestrial carbon-cycle feedback to climate warming: experimental evidence on plant regulation and impacts of biofuel feedstock harvest. Global Change Biology Bioenergy, 1 1 62 74

47. B. E. Medlyn, et al. 2001Stomatal conductance of forest species after long-term exposure to elevated CO2 concentration: a synthesis. New Phytologist, 149 2 247 264

48. Melillo, J. M. et al. (2002). Soil warming and carbon-cycle feedbacks to the climate system. Science, 298, 5601, 2173-21765601 2173 2176

49. S. L. Niu, R. A. Sherry, X. H. Zhou, S. Q. Wan, Y. Q. Luo, 2010Nitrogen regulation of the climate-carbon feedback: evidence from a long-term global change experiment. Ecology, 91 11 3261 3273

50. R. J. Norby, J. Ledford, C. D. Reilly, N. E. Miller, E. G. O'Neill, 2004Fine-root production dominates response of a deciduous forest to atmospheric CO2 enrichment. Proceedings of the National

Academy of Sciences of the United States of America, 101 26 9689 9693

51. R. J. Norby, S. D. Wullschleger, C. A. Gunderson, D. W. Johnson, R. Ceulemans, 1999Tree responses to rising CO2 in field experiments: implications for the future forest. Plant Cell and Environment, 22 6 683 714

52. R. S. Nowak, D. S. Ellsworth, S. D. Smith, 2004Functional responses of plants to elevated atmospheric CO2- do photosynthetic and productivity data from FACE experiments support early predictions? New Phytologist, 162 2 253 280

53. C. E. Owensby, J. M. Ham, L. M. Auen, 2006Fluxes of CO2 from grazed and ungrazed tallgrass prairie. Rangeland Ecology & Management, 59 2 111 127

54. K. S. Pregitzer, A. J. Burton, D. R. Zak, A. F. Talhelm, 2008Simulated chronic nitrogen deposition increases carbon storage in Northern Temperate forests. Global Change Biology, 14 1 142 153

55. P. B. Reich, B. A. Hungate, Y. Q. Luo, 2006Carbon-nitrogen interactions in terrestrial ecosystems in response to rising atmospheric carbon dioxide. Annual Review of Ecology Evolution and Systematics, 37 611 636

56. M. C. Rillig, 2004Arbuscular mycorrhizae and terrestrial ecosystem processes. Ecology Letters, 7 8 740 754

57. A. E. Russell, C. A. Cambardella, D. A. Laird, D. B. Jaynes, D. W. Meek, 2009Nitrogen fertilizer effects on soil carbon balances in Midwestern US agricultural systems. Ecological Applications, 19 5 1102 1113

58. L. E. Rustad, et al. 2001A meta-analysis of the response of soil respiration, net nitrogen mineralization, and aboveground plant growth to experimental ecosystem warming. Oecologia, 126 4 543 562

59. M. A. Sanderson, 2008Upland switchgrass yield, nutritive value, and soil carbon changes under grazing and clipping. Agronomy Journal, 100 3 510 516

60. F. Sartori, R. Lal, M. H. Ebinger, D. J. Parrish, 2006Potential soil carbon sequestration and CO2 offset by dedicated energy crops in the USA. Critical Reviews in Plant Sciences, 25 5 441 472

61. W. H. Schlesinger, J. F. Reynolds, G. L. Cunningham, L. F. Huenneke, W. M. Jarrell, R. A. Virginia, W. G. Whitford, 1990Biological feedbacks in global desertification. Science, 247 4946 1043 1048

62. G. R. Shaver, et al. 2000Global warming and terrestrial ecosystems: A conceptual framework for analysis. Bioscience, 50 10 871 882

63. T. Tans, 2011Trends in atmospheric carbon dioxide." from http://www.esrl.noaa.gov/gmd/ccgg/trends/global.html.

64. D. Tilman, J. Hill, C. Lehman, 2006Carbon-negative biofuels from low-input high-diversity grassland biomass. Science, 314 5805 1598 1600

65. S. E. Trumbore, 1997Potential responses of soil organic carbon to global environmental change. Proceedings of the National Academy of Sciences of the United States of America, 94 16 8284 8291

66. P. M. Vitousek, R. W. Howarth, 1991Nitrogen limitation on land and in the sea-how can it occur. Biogeochemistry, 13 2 87 115

67. L. Vivanco, A. T. Austin, 2006Intrinsic effects of species on leaf litter and root decomposition: a comparison of temperate grasses from North and South America. Oecologia, 150 1 97 107

68. D. A. N. Wang, D. S. Lebauer, M. C. Dietze, 2010A quantitative review comparing the yield of switchgrass in monocultures and mixtures in relation to climate and management factors. GCB Bioenergy, 2 1 16 25

69. J. F. Weltzin, et al. 2003Assessing the response of terrestrial ecosystems to potential changes in precipitation. BioScience, 53 10 941 952

70. W. W. Wilhelm, J. M. E. Johnson, D. L. Karlen, D. T. Lightle, 2007Corn stover to sustain soil organic carbon further constrains biomass supply. Agronomy Journal, 99 6 1665 1667

71. J. Y. Xia, S. Q. Wan, 2008Global response patterns of terrestrial plant species to nitrogen addition. New Phytologist, 179 2 428 439

72. Xue, X.; Luo, Y.; Zhou, X.; Sherry, R. & Jia, X. (2011). Climate warming increases soil erosion, carbon and nitrogen loss with biofuel feedstock harvest in tallgrass prairie. GCB Bioenergy, DOI: 10.1111/j.1757-1707.2010.01071.x

73. D. R. Zak, K. S. Pregitzer, J. S. King, W. E. Holmes, 2000Elevated atmospheric CO2, fine roots and the response of soil

microorganisms: a review and hypothesis. New Phytologist, 147 1 201 222

74. X. Zhou, M. Talley, Y. Luo, 2009Biomass, litter, and soil respiration along a precipitation gradient in southern great plains, USA. Ecosystems, 12 8 1369 1380

75. X. Zhou, S. Q. Wan, Y. Q. Luo, 2007Source components and interannual variability of soil CO2 efflux under experimental warming and clipping in a grassland ecosystem. Global Change Biology, 13 4 761 775

76. V. H. D. Zuazo, C. R. R. Pleguezuelo, 2008Soil-erosion and runoff prevention by plant covers. A review. Agronomy for Sustainable Development, 28 1 65 86

INDEX